Farm Power and Machinery

Farm Power and Machinery

by

Seena V.
Anuja Vijayan

Wordpen Academics
Publishing Science That Matter

ISBN: 978-93-6084-403-5 (Hardback)

Published by : **WORDPEN ACADEMICS**
4736/23, Ansari Road, Darya Ganj,
New Delhi – 110 002
Phone: +91-011-35004336
E-mail: info@wordpenacademics.com

Printed at **:** **Replika Press Pvt. Ltd.**

PREFACE

Farm Power and Machinery is a comprehensive guide that explores the essential role of agricultural machinery and mechanization in modern farming practices. With the continuous advancement of technology and the growing demands for increased agricultural productivity, the need for efficient farm power and machinery has never been more critical. This book aims to provide a detailed understanding of the principles, design, and application of farm machinery, offering insights into how these technologies are revolutionizing agriculture.

Designed for students, researchers, and professionals in agricultural engineering, farm management, and related fields, this book covers both the theoretical foundations and practical applications of farm power and machinery. From tractors and plows to modern automated systems, we explore the diverse range of machines used in planting, irrigation, harvesting, and post-harvest processing. Each chapter highlights the functionality, efficiency, and economic implications of farm machinery, along with the various technologies that drive these innovations.

In addition to the technical aspects, we also discuss the sustainability challenges and the role of mechanization in promoting environmentally friendly and resource-efficient farming practices. As the agricultural sector faces issues such as labor shortages, climate change, and the need for more sustainable practices, the importance of adopting appropriate machinery becomes clear. This book provides a thorough analysis of how farm machinery can address these challenges, from energy-efficient power sources to the integration of precision farming technologies.

We also delve into key considerations such as machinery selection, maintenance, cost analysis, and the impact of automation and robotics on future farming practices. By combining practical knowledge with emerging trends, this book equips readers with the tools necessary to understand, select, and implement farm power and machinery systems that optimize productivity and sustainability.

We are grateful to the agricultural engineers, researchers, and practitioners whose work has contributed to the development of farm power and machinery technologies. It is our hope that *Farm Power and Machinery* will serve as an indispensable resource for those looking to improve agricultural practices through the use of mechanization, enhancing both efficiency and sustainability in the modern farming landscape.

Seena V.

Anuja Vijayan

CONTENTS

	PREFACE	**V**
1.	**ENERGY FOR AGRICULTURE**	**1**
	• Human power	1
	• Animal power	2
	• Renewable energy	4
	• Force, work, power and energy	5
	• Agriculture & Energy Resources	5
	• Water	7
	• Production	9
	• Irrigation	11
	• Fuels for Combustion Engines	12
	• Dry Storage	13
	• Processing	16
	• Tools for Technical Design and Economic Assessment	20
	• Conclusions and Recommendations	21
	• Geothermal	23
2.	**AGRICULTURAL MACHINERY**	**27**
	• The Industrial Revolution	27
	• Agricultural Machinery Types	28
	• New Technology and the Future	29
	• Agricultural Machinery Industry	30
	• Agricultural Machinery Manufacturers	31
	• Agricultural Machinery Industry in Europe	32
	• Mechanised Agriculture	33
	• Productivity-Improving Technologies	37
	• Energy Efficiency	40
	• Electrification and the Pre-Electric Transmission of Power	41

- Mechanization 43
- Machine Tools 45
- Palletization 47
- Field cultivator 49
- Wheel tractor-scrapper 49
- Machinery used for fertilizing & pest control 51
- Crop Harvester 52

3. INTERNAL COMBUSTION ENGINES 55

- External Combustion Engine 55
- Classification of Internal Combustion Engine (IC engines) 56
- Engine Components 60

4. FOUR STROKE AND TWO STROKE ENGINE 67

- Principle of I.C. Engine 67
- Working of I.C. Engine 67

5. SPARK AND COMPRESSION IGNITION ENGINES 75

- Spark ignition engine (Petrol engine) 75

6. COOLING SYSTEM 77

- Purpose of cooling 77
- Methods of cooling the engine 78
- Disadvantages of air cooled engine 79
- Cooling system troubles 83

7. LUBRICATION SYSTEM 87

- Purpose of lubrication system 87
- Types of Lubricants 87
- Flash point and Fire point 88
- Parts of lubrication system Oil pum 91
- Care and maintenance of lubrication system 92

8. POWER TRANSMISSION SYSTEM 93

- Functions of power transmission system 93
- Clutch 94
- Final drive 96

9. ENGINE PERFORMANCE 97

- Brake dynamometer 99

10. TRACTORS AND POWER TILLERS 103

- Tractor Development 103

- Classification of tractor 104
- Components of a tractor 106
- Selection of Tractor 106
- Power take-off unit (PTO) 108
- Operation 111

11. ELECTRIC MOTORS 115

- D.C. Motor 115
- A.C. Motors 116
- Selection of Electric Motors 118
- Care of Electric Motors 119

12. TILLAGE 121

- Objective of Tillage 121
- Ploughing of Land 122

13. PRIMARY TILLAGE IMPLEMENTS 127

- Indigenous plough 127
- General purpose 130
- Slat type 131
- Plough accessories 132
- Vertical suction (Vertical clearance) 133
- Horizontal clevis 135
- Trailed type implement 135
- Terms used in ploughing 136

14. DISC PLOUGH 139

- Advantages of disc plough 139
- Disadvantages of disc plough 140
- Types of Disc Plough 140
- Draft of disc plough 141

15. SECONDARY TILLAGE IMPLEMENTS 145

- Secondary tillage 145
- Spike tooth harrow 150

16. CULTIVATORS, LEVELERS, RIDGER AND BUND FORMER 153

- Cultivator 153
- Cultivator with spring loaded tines 155
- Cultivator with rigid tines 156
- Levelers 158
- Ridger 159

17. SOWING EQUIPMENT **161**

- Methods of sowing 161
- Seed drill 162
- Frame 164
- Furrow openers 166
- PLANTER 168

18. CROP PROPAGATION TOOLS, CUTTING, GRAFTING, PRUNING TOOLS PLANT PROPAGATION **171**

- Reasons for Grafting and Budding 172
- Pruning tools 173

19. INTER-CULTURE EQUIPMENT **175**

- Dry land weeder 176
- Engine operated weeder 177
- Sweep 178
- Junior hoe 178

20. HARVESTING EQUIPMENTS **183**

- Potato Harvester 184
- Tapioca harvester 185
- Turmeric harvester 185
- Fruit harvester 185
- Hoists 187

21. MODERN TRACTORS **189**

- HISTORY 189
- Safety 190
- Applications And Variations 191
- History of Tractors 193
- Types of Tractors 195
- Front Loader 196
- EPA Tractor 197
- Tractor Hazards and Safety 197

22. ECONOMICS OF FARMING **205**

- Agricultural Machinery 205
- Impact of the Industrial Revolution on Agriculture 207
- Agricultural Machinery Industry in India 208
- Farm Power in Agriculture 210
- Farm Machinery Industries in India 212

- Future Thrusts in Agricultural Mechanization 214
- Future Perspective in Agricultural Machinery Manufacture 215
- Study Design and Data 219
- External Economics of Scale 222

23. PRECISION AGRICULTURE 227

- Highlights of Precision Agriculture 228
- Precision Agronomics 230
- History 231
- Importance of Precision Agriculture 234
- Economic and Environmental Impacts 236
- Why Did the Precision Agriculture Important 237
- Drones and Satellite Imagery 238
- The Basics of Precision Farming 240
- Tools for Implementation of Precision Agriculture 246
- Te Emergence and Current Status of Precision Agriculture 252
- Tools to Analyze Data 256
- Precision Agriculture: Crop Condition 259
- Precision Agriculture for Grain Production 261
- Silo storage 262
- Insect control 265
- Growing Corn Using Precision Agriculture 266
- Yield monitor 267
- Harnessing the market for lasting change 268
- Environmental, regulatory issues 270
- From local data to global impact 273
- Precision farming could help reduce climate gas emissions 275
- Effect of precision farming technologies on climate 275
- CAPIGI pays special attention to climate smart farming 276
- Software as a Farm Input 278
- Open Source Software, and Farming 278
- Technology Transition 279

BIBLIOGRAPHY 281

INDEX 285

1

ENERGY FOR AGRICULTURE

Modern agriculture needs modern energy - the two are closely linked. For many developing countries, agriculture is the dominant sector in developing the economy. Increasing productivity and the modernisation of agricultural production systems are the primary drivers of global poverty reduction and energy plays a key role in achieving this. Energy input to modern and sustainable agricultural production and processing systems is a key factor in moving beyond subsistence farming towards food security, added value in rural areas and expansion into new agricultural markets. In many cases, renewable energy technologies and hybrid systems can provide energy services that neatly support the production process, e.g. by providing irrigation (pumps) or post harvest treatment (cooling) or processing (drying, milling, pressing). The requirements of mechanical energy in the agricultural production process are also of critical importance and include human and animal labour as well as fuels for mechanisation, pumping and other activities, and indirectly the production of fertilisers and agrochemicals.

Energy is the primary and most universal measure of all kinds of work by human beings and nature. Everything what happens in the world is the expression of flow of energy in one of it's form. There are a number of different forms of energy available in nature. The different forms of energy used for agriculture (farm power) in India are Human energy (human power), Animal energy (animal power), Mechanical energy (mechanical power such as Tractors, Power tillers, Self propelled combines, Oil engines), Electrical energy (electrical power), and Renewable energy (Biomass, Solar energy and Wind energy).

HUMAN POWER

Human power is the main source of energy for operating small implements and tools at the farm. Stationary work like chaff cutting, water lifting, threshing, winnowing etc. and field work like weeding, broadcasting are also done by manual labour. An average man can develop maximum power of about 0.1 hp and women labour can develop 0.05

hp power for doing farm work. But now a day the availability of human power for farm operations is depleting. The human power working in the field is shown in Plate 1.1.

The advantages of human power are; it is easily available and can be used for all types of work. The limitations of human energy are the costliest power compared to all other forms of power, very low efficiency, requires full maintenance when not in use and affected by weather condition and seasons.

Fig. 1.1 A view of human labour working in the field

ANIMAL POWER

The average power developed by a pair of bullocks is about 1 hp for farm operations. Bullocks are employed for all types farm work in all seasons. Besides bullocks and buffaloes other animals like camels, horses, donkeys, mules and elephants are also used at some places. The average force a draft animal can exert for farm work is nearly 1/10th of its body weight. A pair of bullocks working in the field is shown in Plate 1.2.

The advantages of animal power are easily available, used for all types of work, low initial investment, supplies manures to the field and fuels to farmers and lives on farm produce. The disadvantages of animal power; viz. they are not very efficient, seasons and weather affect the efficiency, cannot work continuously, requires full maintenance when not in use and very slow in doing work.

Fig. 1.2 A pair of bullocks working in the field Mechanical power

Broadly speaking, mechanical power includes stationary oil engines, tractors and power tillers. The internal combustion engine is a good device for converting fuel energy (Chemical energy) into useful work (Mechanical energy). The internal combustion engines are of two types : Diesel engine and Petrol / Kerosene engine. The thermal efficiency of diesel engine varies from 32 to 38 per cent whereas that of petrol engine varies from 25 to 32 per cent.

In modern days, almost all the tractors and power tillers are operated by diesel engines. A tractor in field operation is shown in Plate 1.3. The diesel engines are used for stationery operations like water lifting, flour mills, oil expellers, vegetable washers, gardeners, mulberry processing equipments, threshers, winnowers, chaff cutter etc.

The advantages of mechanical power are; their efficiency is high, not affected by weather, can run at a stretch for longer period, require less space and cheaper source of power. The disadvantages of mechanical power are; initial capital investment is very high, fuel is costly and scarce, needs technical knowledge for repairs and maintenance.

Fig. 1.3 Mechanical power – Tractor in operation Electrical power

Electrical power is the mostly used for operating electrical motors in the farms. The different types of electric motors used are shown in Plate 1.4. This is a clean and efficient source of energy. The maintenance and operation of motors needs less attention and care. The operating cost remains almost constant throughout its life. Electrical power is used for operating pump sets, high-tech nursery, diary industry, cold storage, farm product processing, fruit industry and food processing industries.

The advantages of electrical power are viz; very cheapest form of power, high efficiency, can work at a stretch continuously for longer periods, maintenance and operating cost is very low and not affected by seasons. The disadvantages are viz; initial capital investment high, requires good amount of technical knowledge and if handled carelessly, it causes great danger.

Fig. 1.4 a

FIg. 1.4 b Plate : 1.4 Different types of electric motors

(Source: http://www.indiamart.com)

RENEWABLE ENERGY

It is the energy mainly obtained from renewable sources like sun, wind and biomass. Biomass energy (biogas, producer gas, ethanol and biodiesel), wind energy and solar energy are used in agriculture and domestic purposes with suitable devices. It can be used for lighting, cooking, water heating, space heating, water distillation, food processing, water pumping, diesel engine operation and electricity generation. This type of energy is inexhaustible in nature. The different renewable energy gadgets are depicted in Plate 1.5.

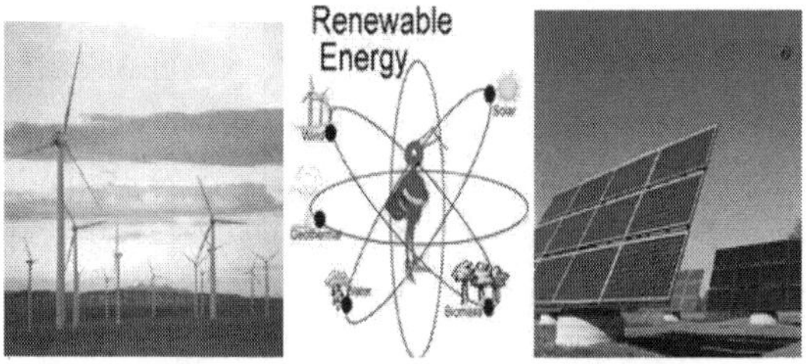

Fig. 1.5 Different renewable energy gadgets

(Source: http://www.treehugger.com , http://greenerbuyer.com & http:// c1.ecopreneurist.com)

FORCE, WORK, POWER AND ENERGY

Force : An invisible agent which always tries to change the state of the body. The unit for force in SI system is Newton (N) and in MKS system is Kg-f

Work : When a force is applied to the body, it moves in the direction of the applied force, then work is said to be done. The unit for work is Newton- meters (N-m).

Work done = Force X Distance traveled N-m

Power : It is the rate of doing work. The work done per unit time is called power and is expressed in N-m-Sec-1.

$$\text{Power} = \frac{\text{Work done}}{\text{Time taken}} \quad \text{N-m-Sec-1 or Watt}$$

Energy: It is the capacity to do the work. The unit for energy is Joules or Calories or Kilo Watt or Horse Power.

Horse Power:- It is that amount of force which is capable or displacing 75 kg force through a distance of one meter in one second.

AGRICULTURE & ENERGY RESOURCES

Agricultural production more than tripled between 1960 and 2015, owing in part to productivity-enhancing Green Revolution technologies and a significant expansion in the use of land, water and other natural resources for agricultural purposes. The same period witnessed a remarkable process of industrialization and globalization of food and agriculture.

Food supply chains have lengthened dramatically as the physical distance from farm to plate has increased; the consumption of processed, packaged and prepared foods has increased in all but the most isolated rural communities.

Looking ahead, the core question is whether today's agriculture and food systems are capable of meeting the needs of a global population that is projected to reach more than 9 billion by mid-century and may peak at more than 11 billion by the end of the century. Can we achieve the required production increases, even as the pressures on already scarce land and water resources and the negative impacts of climate change intensify?

We need food and clean fresh water to produce food. Due to the growing demand, on the one hand, and climate change on the other, many regions, particularly in the South, have difficulties to find the amount of freshwater needed to fill their needs.

How can we continue to cultivate without taking into account nature? More efficient use of water in agriculture would certainly help.

But water is not the only point that we have to improve in the agricultural sector. Agriculture is responsible for a big part of energy consumption. Issues in this sector are very important if we want to reach our goals, in the field of climate change, but humanity already has some solutions in his hands.

Energy

From the production of fertilizers to the processing and transporting of food products to market, the industrial food system depends on fossil fuels to produce monocultures of commodity crops. Increasingly, food crops are being diverted to energy production — most notably corn, which is being used to make ethanol as a fuel. Industrial agriculture's present reliance on finite energy sources that negatively impact the natural environment is not sustainable. There are energy alternatives at every step along the process that can help our food system become more resilient. 80% of energy needs in the agricultural sector are satisfied by fossil energies.

The consumption of petroleum products (domestic fuel oil and heavy fuel oil, diesel and gasoline) accounts for almost 70% of the total energy consumption of the agricultural sector in 2011.

Natural gas consumption still accounts for only 8% of this consumption. Even if it has grown strongly in this sector since 1980. The share of thermal renewable energies and the recovery of waste is still modest, only 4% of agricultural energy consumption in 2011 was from renewable energies.

Processing, Packaging and Transportation

Much of the food produced today is highly processed and heavily packaged, which further increases its energy footprint. In creating processed food, energy is used to grind, chop and mix products together and to cook, bake or otherwise handle the product. Manufacturing food packaging also requires energy; most food packaging is made from plastic, which is made out of petroleum. The amount of energy required for packaging varies greatly depending on the size, type and quantity of the materials. As a result of consolidation and centralization of production, foods are often transported long distances, requiring additional energy inputs.

Energy and Animal Agriculture

Most meat, eggs and dairy products are now produced on factory farms, huge industrial livestock operations that raise thousands of animals in confined conditions without access to pasture. Since the animals are unable to graze, factory farms require tremendous quantities of feed produced by industrial crop farms using the energy-intensive processes described above. About half of the 90 million acres of land used to grow corn ends up as feed grain. And corn makes up about 95 percent of feed grain production. Out of all acres of harvested cropland in the US, corn represents about 28 percent. Factory farms are also potential sources of ground and surface water pollution, which may require municipalities and treatment facilities to expend additional energy and resources on water treatment.

These sources of energy are mainly used for heating systems, to warm indoor livestocks.

We consider that for a cow we will need 20 litres of fuel to grow it until it get killed, 3 for a chicken and 33 for a pork, only for the heating system. We have to add to these spendings the consumption of fuel for transportation and to grow their food.

The use of pesticides, fertilizers and herbicides represent also a big part in the energy consumption of agriculture. In the United States, for instance, to spread their products, they rent planes which are very polluting and harmful for the environment.

The remaining 18% of the energy consumption of the agricultural sector is supplied with electricity (mainly produced by nuclear reactors in Europe). In 2011, electricity consumption in agriculture reached twice as much as in 1985 in Europe.

But we have to notice changes in this area. In the Netherlands, the agricultural world is one of the most innovative sectors and this is very interesting when we consider energy spendings.

Of course, not all innovations are technological, and a number of them are based on know-how and real knowledge of nature. Some techniques rely on nature in itself. The creation of "smart" ecosystems involving species of insects, mites and microscopic worms to fight against harmful insects seems to work very well, to replace herbicides or pesticides for example. This is the principle of "biological control", which protects crops. This biological control is one example among many other, but seems to be a real solution for the future if we want to erase pesticides pollution and if we want to find an alternative for GMOs.

WATER

Water used in agriculture can have several origins. Green water comes from rainfall and is stored temporarily in the soil. Blue waters are pumped into rivers, lakes or groundwater, but in a sustainable way. The third source corresponds to the water extracted from reserves which renew themselves little through time or not at all.

Unfortunately a lot of farmers tend to use unsustainable water, extracted legally or sometimes illegally. Most of the time, this water is used for drinking water for animals or crops irrigation.

Irrigation, which was already used by Egyptian and Mesopotamian civilizations, is still a developing technique today. It is often used in addition to rain-fed irrigation to increase yields and increase the length of the agricultural season. Still very often traditional, irrigation devices have a very low yield. It is estimated that with standard techniques, 30 to 60% of irrigation water is lost or evaporate, and do not benefit crops.

Researchers from the University of Utrecht, led by Marc Bierkens, decided to study the source of irrigation water used in the world. Their results, published in Water Resources Research, are disturbing. The use of water resources without real sustainable management tripled between 1960 and 2000, from 75 to 234 km3 per year. It means that natural water reserves, without a sustainable management, would be definitely out of use in few decades.

- 70% of the worldwide water consumption is for agriculture. This number is insane when we know that almost 30% of humans don't have access to drinking water, which is approximately 2.1 billion people.
- In Europe, only 35% of the water consumption is for the agricultural sector, even if it is a big part, it is less than Asia where 83% of the water is used for agriculture.
- In the agricultural sector, we can find two types of water use. The first one is directly for human consumption and the other one used for animal foods or animal drinking water. Livestock related activities account for at least 80% of the US water consumption, almost 95% of agricultural consumption.
- More surprisingly, some countries resorting mainly to blue waters (lakes and rivers) and green (rain) could end up in water deficit due to global warming (decrease in rainfall). Algeria, Morocco, Spain and Italy are particularly concerned.
- In the hypothesis of water decrease, the economic consequences would be felt on a global scale. Netherlands import huge quantities of fruits and vegetables from Spain and Morocco, two countries mentioned.
- A decrease in global agricultural production could also have serious health consequences, as the population continues to grow. Approximately 9 billions people will live on earth in the next decades and this population will consume on average 50% more than the actual population.
- The consumption of water has to be reduced, if not, we will face serious production problems in the following years.

The meat

At the crux of our inefficient use of land, water and soil is the fact that, currently, over 50% of the world's crops are used to feed animals and not people. This problem will certainly be exacerbated in the future given that, increasingly, consumption of animal products correlates strongly with both population growth and economic growth. Further compounding this pressure on food production is the increasing demand for energy that is encouraging more bio-fuel production from edible crops like corn and sugar. These forces are converging in a world where more than 1 billion people are undernourished.

Land

At present some 11 percent (1.5 billion ha) of the globe's land surface (13.4 billion ha) is used in crop production (arable land and land under permanent crops). This area represents slightly over a third (36 percent) of the land estimated to be to some degree suitable for crop production.

Agricultural land (including natural grassland) accounts for almost half of the European territory (48%).

Deforestation is the removal of trees from the land by man-made and natural events. Each year, 115 to 150 million square kilometers of forest are lost due to deforestation. According to the World Bank between 1990 and 2016, the world lost 1.3 million square kilometers of forest.

In the last 40 years we have lost 50% of all rain forests and over time we have lost half of all forests globally. Deforestation is mainly caused by agriculture (80%), logging (14%), and firewood (5%).

The biggest cause of forest loss – accounting for around 80 per cent – is agricultural deforestation, notably for beef, soy, palm oil and commercial timber. Soy ranks as the second largest agricultural driver of deforestation after cattle products.

The EU is involved mainly because European farmers rely on imported soy from Brazil, Argentina and Paraguay to use as animal feed. Of all the soy meal consumed in the EU, 83 per cent goes in concentrate feed for pigs and poultry.

PRODUCTION

Land Preparation, Cultivation, Harvesting and Threshing

The power necessary for agricultural production is provided by human labour, draught animals or engine-driven machinery. Mechanisation is a key input in any farming system applying tools, implements and machinery to improve the productivity of farm labour and land. Based on the power source, the technological levels of mechanisation have been broadly classified as hand-tool technology, draught animal technology and mechanical power technology.

The bulk of direct energy inputs in smallholder production is provided by human and draught animal power (DAP). Agriculture in developing countries relies heavily on the physical capability of farmers, with often limited output, depending on the physical energy available. A fit person consumes around 250-300 W in terms of energy, depending on climate and with a rest of 10-30 minutes/hour. The efficiency of energy conversion is only 25 per cent, with a maximum power output of 75 W. Operations involving human skill include activities such as planting, weeding, spraying and harvesting and using hand tools. Most horticultural commodities are harvested manually, including those grown in highly industrialised nations. Energy requirements during harvesting include mechanical energy for digging up root and tuber crops. For example, a simple, labour-saving cassava harvesting tool was introduced to Thai farmers by the International Centre for Tropical Agriculture (CIAT) reducing person-days from 15-40 (depending on the type of soil and the yield) to 5-10 days only. The tool costs 2.50 USD, is produced by local manufacturers and can be used collectively by farmers. Harvesting early in the morning when air temperatures are cooler also helps to reduce energy requirements and cooling costs.

Activities such as ploughing, soil preparation, water lifting, pulling inputs and threshing require fewer skills but greater energy input and are hence mostly powered by draught animals such as buffalos, horses, donkeys, camels or oxen. Power output ranges from 200 W for a donkey for 4 hours daily work to over 500 W for a buffalo . Draught animal power can alleviate human drudgery and is generally considered to be an affordable and sustainable source of power for smallholders, also given its great potential for diversification and expansion e.g. for transport and non-farm work. Efforts to expand the use of draught animal power should include work on animal efficiency, which can be improved through modernisation of equipment, better breeding and animal husbandry, feeding and veterinary care.

In sub-Saharan Africa (SSA), human effort contributes about 65 per cent of the power required for land preparation and subsequent weeding, the principal demand peaks in the farming cycle. An example comparison shows that a typical farm family relying solely on human power can cultivate only 1.5 hectares per year. This rises to 4 hectares if animal power is available and to over 8 hectares if tractor power can be accessed. Transition to increased mechanisation of agricultural operations occurs using diesel or gasoline-powered tractors and harvesters. Fuel powered two-wheel tractors are often used for agricultural production, with different attachements for tillage, bed planting, row planting, harvesting and threshing. In Bangladesh, power tillers (12-15 hp) are used for about 70 per cent of farm work because of their versatility. The fuel requirements for digging up root and tuber crops, e.g. of a potato digger are 0.57 gallons of diesel fuel per ton of product (1.96 litres/MT). Energy use depends mostly on the speed at which activities are carried out. As tillage operations within arable farming systems often have the highest energy requirements and lead to soil degradation, the reduction of mechanical tillage is promoted by the approach of conservation agriculture (CA). In general, a four-wheel tractor will not be economically feasible for a smallholder with a typical land holding of up to 5 hectares. Government-run tractor hire schemes in SSA countries have not been able to increase farm productivity. On the other hand, the concept of a rental market for privately owned and operated tractors has potential that may increase in the future .

Besides direct energy demand in agriculture there are also indirect energy inputs in the form of sequestered energy in fertilisers, herbicides, pesticides, and insecticides, all of which require energy in their production, distribution and transport processes. Fertilisers form the largest of these energy inputs to agriculture, whilst pesticides are the most energy-intensive agricultural input (on a per kg basis of chemical) . Need for these inputs however can be reduced by the use of farm-own organic manure or by agro-industry based bio-fertiliser production, e.g. generated from biogas plants (at household or industrial level). The promotion of locally produced organic fertiliser also has the advantage of avoiding the negative environmental impact of mineral fertilisers and benefits from the shorter distances required for distribution, thus reducing energy requirements and fuel costs.

Past experience in Asia indicates that the key factors driving mechanisation and consequently power for agricultural development are guaranteed markets and farm prices for key crops, favorable government policies and availability of credit. The example of India shows that favorable government policy and farmer demand have led to increased availability of appropriate and locally manufactured farm machinery and ultimately to India becoming the world's largest tractor manufacturer. A very comprehensive comparison of Energy Use in Conventional and Organic Cropping Systems has been carried out by the National Sustainable Agriculture Information Service (ATTRA) of the US National Centre for Appropriate Technology (NCAT). The report shows, in a review of existing studies, that there are many complexities involved in comparing energy consumption in conventional and organic cropping systems. Certain research indicates that organic agriculture is more energy-efficient than conventional agriculture, but not in all cases. In some cases, organic agriculture may be more energy-intensive, depending on the specific farming operation, the crop produced and the post-production handling. It is important to assess the energy-intensiveness of food systems in a holistic manner that incorporates energy consumption for the entire life cycle of the food product.

IRRIGATION

Access to water is a major determinant of land productivity and active water management through irrigation offers an important opportunity to stabilise yields. Irrigation is therefore of the utmost significance in the agricultural production process and this chapter is consequently accorded great importance. Irrigated land productivity is more than double that of rainfed land. In sub-Saharan Africa, only 4 per cent of the area in production is under irrigation, compared with 39 per cent in South Asia and 29 per cent in East Asia. With growing water scarcity and costs of large-scale irrigation schemes rising, there is a need to enhance productivity by improving existing schemes, expanding small-scale schemes and developing water harvesting. Irrigation refers to the distribution of water for growing crops, including the use of water storage and pumping, where appropriate. The energy demand for irrigation purposes is the energy required to lift water by pumping from surface sources, such as ponds, streams, or canals; or from below-ground sources using open wells or boreholes. This water is typically pumped to surface canals, reservoirs, or elevated tanks. The energy demand for water lifting is calculated by multiplying the head (the vertical distance from the water source to the field in metres) by the volume of water to be raised in cubic metres (m3). Pumping energy needs are typically expressed in units of m4.

Decisions about energy and irrigation should be made on the basis of three key sequential steps :

1. Choosing the right irrigation technology: understanding the irrigation system needs for a given context,
2. Choosing the right pump technology: understanding the various pumping technologies available,
3. Choosing the right energy source: assessing which power sources are possible, desirable, and locally available to provide the necessary energy for that system.

FUELS FOR COMBUSTION ENGINES

Motorised pumps can be powered by fossil fuels (diesel, gasoline), either through generators that create electricity, or by transmitting power to the pump through a drive belt and vertical rotating shaft. In addition, some submersible pumps (i.e. progressing cavity pumps) operate by direct displacement, like piston pumps. The pumps tend to be more expensive but nevertheless more efficient than centrifugal pumps. The use of biofuels (plantoil, biogas) is still limited. A few experiments are underway, e.g. in India with biofuels made from local oilseeds , in the GIZ PSDA program in Kenya with biogas and in Nepal, where Winrock International is supporting community-grown jatropha oil for irrigation pumps .

Wind Energy

Wind can be used to power both mechanical and electric pumps. Mechanical wind-powered pumps use reciprocal non-motorised submersible pumps and require wind speeds of 2.5 m/s minimum up to 4 m/s optimum. Capacity is much lower than for motorised centrifugal pumps, in the range of 1 m3/hr at depths of 20 metres or more.

Mechanical wind pumps require the availability of local maintenance and repair facilities to be able to respond quickly to mechanical failures. Adequate wind speeds must be present at the location of the wells. One advantage is that they can pump day or night as long as there is sufficient wind, and can be used independently of electricity or fuel supplies. A disadvantage is that they must be located directly above the well, a location that may not be optimal in terms of local wind resources. Wind pumps are appropriate in windy areas without other sources of power, and only for small irrigable areas.

Wind electric turbines convert the kinetic energy of the wind into rotational mechanical energy that drives a generator to produce electricity, e.g. for pumping water for irrigation. Windmills are positioned for optimal wind conditions, providing greater site flexibility and in addition facilitating electricity production for other uses. Water-pumping applications generally make use of wind turbines with rated output between 1 kWe and 10 kWe. A wide variety of small wind electric turbines is commercially available, with rated outputs ranging from a few tens of watts to 100 kilowatts, and is used worldwide to provide electricity in locations where alternatives are unavailable or are too expensive or difficult to provide.

Hybrid Systems

Small-scale hybrid power systems, also a mature technology, are used worldwide. By combining different energy sources (solar-diesel, wind-diesel) hybrids can provide widespread and highly reliable electrical supply. These small hybrid systems are easy to install—no special tools or concrete are required.

Post-harvest and Storage

Packing

The best precondition for storage is a proper harvest and post-harvest handling. Sorting, grading, and packing produce into storage containers after harvest to prevent losses is important, especially for easily perishable crops such as tuber crops, vegetables and fruits. Reducing the number of times produce is handled between harvest and consumption will reduce mechanical damage and subsequent losses. Packing houses for crops are often simple structures that provide shade and comfortable working conditions for farm workers conducting manual post-harvest operations. Manual handling is recommended for horticultural high-value produce.

Storage

A certain percentage of the farm produce has to be stored before selling at the market, because production is seasonal while demand, particularly for vegetables and fruits, continues year-round. Storage facilities secure agricultural commodities kept for rapid emergency aid and buffer stocks to stabilise domestic prices. Besides storage at central hubs and transportation centres, more localised storage is often necessary.

DRY STORAGE

To achieve increased production, traditional village stores need to be improved within the means of smallholders to achieve a certain transition to larger and more modern storage systems. This requires the management of dry storage facilities, mainly for post-harvest and storage of grain (maize, sorghum, millet, wheat and rice), pulses (beans, peas), roots and tuber crops, and oilseeds, whereas fruits and vegetables require cold storage facilities. For grain, pulses and oil seeds, mechanical damage mostly occurs during harvesting, transport and processing, while insects, mold and rodents damage produce during storage. For such staple foods, energy input or expensive technology is generally. Good storage practice is the key to maintaining quality and value throughout the storage season. A precondition is that crops are in good condition prior to storage, which means well cleaned and carefully winnowed to remove live adult insects. Straw, chaff, weed seeds, stones and dirt must also be removed, as they hold water and their removal will allow grain to dry faster. The grain should be well dried to an appropriate level of moisture content. Good drying is essential because damp grain will become moldy and spoil. Solar dryers can be used—sun drying will also help get rid of adult insects in the grain. However, as sun drying will not kill all immature stages, such as larvae living inside grains, it is sometimes necessary to treat the grain with insecticide, thus killing the immature stages once they mature and emerge from the grain as adults. Maize is often shelled by hand. This produces better quality grain but is a very slow process. Quicker and effective shelling can be achieved using a variety of mechanical operated machines. The grain must be sufficiently dry both for these and for safe storage. A good store will keep the grain dry and cool and provide protection against rodents and birds. Practising good store hygiene—keeping everything as clean as is practically possible—helps maintain the condition of the crop and the store throughout the storage season. As pests can attack the store at any time, it is important to inspect the store and crop regularly. There are many appropriate technologies and traditional methods for preventing post-harvest losses, varying from location to location. Dry storage facilities range from small-scale, on-farm storage to medium and large-scale facilities of several hundred of tons or more.

Cold Storage

Cold storage in tropical and subtropical climates can generate high energy demand. In general, the initial cooling, processing and cold storage of fresh fruit and vegetables is among the most energy-intensive activities of the food industry. However, for fresh fruits and vegetables, cooling is also one of the most important steps in the post-harvest handling chain to reduce respiration rates, extend shelf life and protect quality. Deterioration of fruits and vegetables during storage depends largely on temperature, so the temperature must be lowered to an appropriate level, to avoid damage and increase the length of storage time. Controlled temperature storage is a critical factor for most perishable agricultural products and a consistent cold chain is necessary to maintain the quality of many high-value agricultural products. Cooling also offers farmers the opportunity to increase income by extending the period for selling and marketing the products when better prices can be achieved. In developing countries, refrigeration in rural areas must often be operated without reliable electricity supply and the provision of a cold chain or store rooms for cooling bigger quantities of produce is often impossible. Three technologies are available for cooling smaller quantities of produce:

- Passive/evaporative coolers: where temperatures between 10-25°C are needed, no energy input is required.
- Absorption refrigerators (heat driven coolers (HDC), requiring temperatures below 10°C; heat as an energy source is used to drive the cooling system (e.g. solar, kerosene-fueled flame) .
- Refrigerators, which are electrically driven, use mechanical compression technology. Refrigeration is dependent on a reliable and continuous supply of electricity which can be generated from different sources: from the grid, diesel powered compressor, mechanical power generated from a water turbine, by solar energy converted by solar cells into electricity stored in batteries or by solar energy and evaporation.

Commercial PV-powered refrigerators were introduced in developing countries to cut down the use of kerosene or gas-powered absorption refrigerators, which are the most common alternatives but also those with the most negative impact on the environment. First introduced for medical refrigeration they now also constitute an alternative method of storing small quantities of agricultural products for sale. The newest type of solar refrigerator is solar powered but requires neither solar panels nor a battery as it functions on the principle of evaporation. Solar photovoltaic power for refrigerators has great potential for lower running costs, greater reliability and a longer working life than kerosene refrigerators or diesel generators. A life-cycle cost analysis has shown that the costs are approximately the same for solar and kerosene refrigerators, but because they are more reliable and environmentally friendly, solar refrigerators are the preferred option.

Categories	Commodities Technologies	Energy Sources
Low tech (<5 kWh/day)	Field packing of leafy, stem, or fruit vegetables, root, tuber and bulb crops, fruits and berries	Electric grid; Solar power with battery back-up
Basic tech (5 to 25kWh/day)	Packinghouse operations and pre-cooling for tropical and subtropical fruits and vegetables; Evaporative cool storage. (Temperature range 15°C to 20°C)	Solar water heater, Electric grid; Generator (diesel or gas); Hybrid PV/ Generator systems with battery back-up

Intermediate tech (25 to 100 kWh/day)	Cooling and cold storage for temperate fruits and vegetables. (Temperature range 0°C to 7°C)	Electric grid; Generator (diesel or gas)
Modern tech (> 100 kWh/day)	Automated packinghouse operations, pre-cooling and cold storage for any kind of fruits and vegetables. (Temperature range down to 0°C)	Electric grid; diesel back-up generators

PROCESSING

Drying of Produce

To add value to fresh fruits and vegetables, they must be dried before packaging and marketing. Drying of fresh produce, which contains up to 95 per cent water, to a safe moisture content of 7-8 per cent requires the application of low heat and ventilation for best results. Processing fruits and vegetables on a small-scale level is therefore less mechanised than other agro-processing activities. It is important that the used equipment is constructed of stainless steel because acids in fruits corrode mild steel. Peeling, cutting, or slicing produce into uniformly sized pieces is usually required for successful drying. This is typically done by hand. Machines are available for larger quantities and can speed up the peeling, cutting, or slicing of produce. For drying commodities (e.g. tea, coffee, copra or abaca) in larger quantities, heat assisted drying is used, with many different types of dryer available.

Direct and Indirect Solar Drying

Laying produce out in the sun to dry naturally is common in sunny climates and is a very inexpensive drying method. It is still the predominant method for producing raisins worldwide. However, traditional sun drying methods often yield low quality as the produce is not protected against rain and dust and other contaminations.

Indirect solar drying requires a covered dryer that protects produce from direct sunlight while capturing more heat from the sun. With natural air flow inside, an indirect dryer reduces heat and pest damage while speeding up drying.

Heat-assisted Drying

In rainy weather regions or humid locations, heat-assisted dryers create warm air flow inside the dryer to speed up drying. Heat sources may be electric, propane, wood, or any other locally available fuel. Appropriate drying temperatures can easily be achieved with low-powered, relatively inexpensive technologies. The majority of energy used in heated air dryers is actually used to heat the air. Electricity for moving air is only a small fraction of the air heating costs, which depend on the initial and final moisture contents of the product. If energy conservation measures such as air recirculation are incorporated into the operation, fuel use can drop significantly.

Solar Powered Irrigation Systems

Irrigated agriculture is the backbone of local economic development in the majority of developing countries. Reliable and affordable access to irrigation water is a crucial factor as the article on irrigation emphasizes already.

The technological option of Solar Powered Irrigation Systems (SPIS) is rarely taken into consideration due to a lack of pertinent experience and the comparably high investment costs of the past. Nevertheless, as prices for solar modules have fallen substantially in recent years, governments, extension services and technical cooperation are reconsidering photovoltaic water pumps (PVP) to be employed in agricultural production and beyond. Demand in this regard will have to be largely generated from the side of rural farm households themselves.

As indicated by the Powering Agriculture Energy Grand Challenge (PAEGC), some barriers exist, which hinder the integration of "clean" energy technology in agriculture development:

- Farmers and extension services are not aware of the variety of new technologies that may be appropriate for them;
- Clean energy technologies are relatively new, therefore farmers have limited access to distributors for installation, parts, and service;
- Farmers often do not have the means to cover high capital costs associated with clean energy upgrades - and financing is seldom available.

The Deutsche Gesellschaft für Internationale Zusammenarbeit (GIZ) GmbH is implementing BMZ's contribution to this initiative and has commissioned a worldwide study on technical solutions and management practices related to solar-powered irrigation systems (SPIS). The study is developed in close cooperation with the sector program Poverty-oriented Basic Energy Services (HERA), which has more than three decades of experience in the field of energy access.

Overview of Irrigation Technologies

This chapter gives an overview about modern irrigation technologies, their advantages and disadvantages, required energy sources as well as financial aspects:

- General types of irrigation
- Surface irrigation

- Sprinkler irrigation
- Drip irrigation
- Comparison of sprinkler and drip irrigation

1. Micro-sprinkler irrigation
2. Modern water-saving irrigation solutions

- Cropping pattern requirements
- Efficiency and durability
- Energy sources for water abstraction and pressurized irrigation
- General classification and description
- Hybrid solutions
- Pumping technology for water abstraction and irrigation
- Comparative financial analysis of irrigation solutions
- Generic costs of irrigation solutions
- Financial impact on farm budgets
- Financial services requirements
- Financial calculation models
- The report is currently under revision and not available on energypedia.

Technical Characteristics and Design of SPIS

The chapter technical characteristics and design of SPIS introduces the following aspects:

- Components of photovoltaic pumping systems
- Solar generator
- Mounting and solartracking systems
- Controller and motor
- Water pump
- Water storage tank
- Monitoring
- SPIS plant concepts
- Variability of global solar radiation
- SPIS plant configurations and operation
- Suitability for drip irrigation

- Filter systems
- Fertigation systems
- Planning and sizing of SPIS
- Design data collection
- Estimation of PV generator size
- EXCEL-based system sizing
- Computer-based system sizing and simulation
- Land requirements for SPIS
- The report is currently under revision and not available on energypedia.

Management Requirements of SPIS

For assessing the specific management requirements of PV irrigation systems, it is advisable to distinguish between strategic, tactical and operational management:

- Stakeholder in the SPIS management
- Impacts on strategic farm management
- Impacts on tactical farm management
- Impacts on operational farm management
- The report is currently under revision and not available on energypedia.

Financial Viability of SPIS

Financial viability is the ability to generate sufficient income to meet operating expenditure, financing needs and, ideally, to allow profit generation. Each irrigation technology choice has its limitations in terms of the range of crops it can support in production. On the other hand, the investment costs attached to the choice of technology limit the range of crops to select for production to a large extent. Pressurized irrigation systems in general and micro irrigation systems in particular are on the capitalintensive side of investments for irrigation infrastructure and equipment.

This chapter explains respective parameters for assessing financial viabilit, provides an exemplary financial analysis of eight SPIS systems and includes business models for SPIS.

Ecological Impacts and Sustainability of SPIS

Apart from being a cost-competitive and reliable source of electricity, photovoltaic systems generate a number of ecologic benefits. This chapter describes:

- Ecological impacts:
- The Carbon Footprint of photovoltaic systems

- Energy payback time
- Recycling of solar panels
- Reducing the risk of groundwater depletion
- Avoidance of groundwater contamination
- Reducing soil salination
- Avoidance of noise and exhaust fumes emissions
- Sustainability of SPIS:
- Technical reliability of system components
- Detected failures and trouble shooting
- Availability of SPIS components on local markets
- Spare parts and after sales service
- Acceptance of SPIS
- Water governance issues
- The report is currently under revision and not available on energypedia.

TOOLS FOR TECHNICAL DESIGN AND ECONOMIC ASSESSMENT

In this chapter, a selection of the individual software tools will be presented. The information has been compiled from reliable documented and published references/ resources as cited in the publication. The following tools are presented:

- Design and simulation tools for PV pumps
- Design and simulation tools for irrigation systems
- Calculation tools for irrigation requirements
- The report is currently under revision and not available on energypedia.

Country Case Studies

Country case studies were carried out in Chile, India, Kenya and Morocco as part of the stocktaking and analysis exercise. The selection of countries was made to have a worldwide bias and an insight into different markets and market scopes:

- Chile: Widespread utilization of PV pumping technology in irrigation, strong promotion of SPIS by government, main target group small to medium farms (2 - 5 ha), but also larger farms up to 30 ha under PV irrigation;
- India: High level of water abstraction with pumps, ambitious government program to replace conventional pumps by PV technology, main target group small to medium farms (2 - 5 ha), local products/systems on the

market;

- Kenya: Very high share of smallholders (< 2 ha) and high potential to develop dry lands by PV pumping from groundwater, local products on the market;
- Morocco: Main market for suppliers like Lorentz and Grundfos, medium (10 – 15 ha) and large size commercial farms utilizing PV pumping
- The report is currently under revision and not available on energypedia.

Potential and Barriers for SPIS Distribution

Several barriers need to be overcome in developing countries to develop their huge PV potential. This chapter gives a brief overview of key drivers for solar pumps and key hurdles that stand in the way of SPIS dissemination. An opportunity and risk assessment in the following aspects has been conducted:

- Opportunity assessment
- Global PV market growth
- Diesel generator replacement
- Greenfield development
- Technology access to smallholders
- Bridging grid power failures
- Job creation and local production
- Innovation potential
- Assessment of Barriers and Risks
- High initial investment cost
- Lack of market-oriented financing
- Oil price development
- Lack of market-oriented policies
- Lack of awareness and impact monitoring
- Lack of quality assurance and service
- Natural disasters and theft
- Summary of opportunity and risks
- The report is currently under revision and not available on energypedia.

CONCLUSIONS AND RECOMMENDATIONS

The analysis in this report underlines that photovoltaic powered irrigation is a technically mature option, even when it is not yet very widespread. Solar powered water pumping can be built into irrigation systems in different ways. The introduced case studies from different countries give an insight into the wide range of application of the technology. From a technical point of view, photovoltaic water pumping can be integrated into most irrigation concepts. The technological development on the side of the pump manufacturers is far advanced and the market can hence provide a suitable pumping solution for almost any requirement and condition. This includes the integration of PV pumps into hybrid systems.

The information collection and the analysis brought together in this report show that agronomic and financial feasibility requirements limit the range of application of PV technology in irrigation. The promotion of the technology will have to take these limits into account by departing from the fact that the utilization of PV technology requires a high initial capital investment and technological know-how for system design and development.

The absence of suitable financing products catering for the specific needs of SPIS development (high initial capital needs, no additional collateral options, long repayment period) is an obstacle to the dissemination of the technology. Good examples like in India and Morocco show that corresponding loan financing is an option, even though it may require a particular risk management.

Renewable Energy Production on Farms

In combination with energy conservation practices, farmers can produce their own energy to become even more self sufficient by reducing external inputs. Not only does renewable energy help the farmer save money but also combats the effects of global warming. Biomass, geothermal, hydroelectric, solar, and wind power can produce electricity for heating, lighting, and fuel for use on the farm. This publication describes and outlines appropriate uses for the renewable energy options mentioned above in a farm setting, as well as providing some information on finding financial incentives for installing and running these systems.

With a little initial investment, energy can be made for free from renewable sources such as the sun, wind, and water. When considering a renewable energy system for your site check with local zoning ordinances before beginning construction. Also keep in mind that any renewable energy electric system on the farm can be connected to the utility grid with an inverter allowing the farmer to receive payment for the power they produce.

Biomass

According to the Union of Concerned Scientists, "Tripling U.S. use of biomass for energy could provide as much as $20 billion in new income for farmers and rural communities and reduce global warming emissions by the same amount as taking 70 million cars off the road." Plant and animal derived materials used for energy are known as biomass. Biomass such as sugars and oils from plants can be used to make fuel for vehicles (Biofuel or Biodiesel) and the burning of biomass for heat or electricity is simply called Biopower.

Biofuels are a renewable energy alternative which can be made from crops grown on the farm to fuel vehicles. Some of the controversy surrounding biofuels has to do with using valuable agricultural land to grow energy crops rather than food. To combat this controversy, some farmers choose to grow oil fuel crops such as sunflower, canola, or crambe for the dual purpose of feedstock and biodiesel production. Growing oilseed crops for fuel production on the farm can also be cost effective where after crushing the seed to extract oil, the meal byproduct has a higher market value than the oil itself. While small scale oil presses may be expensive (ranging from $400 to $13,000) some farmers have found a cooperative purchase with other farmers to be effective. Some disadvantages of biodiesel are poor cold weather performance and formation of sediments if stored for long periods. Recycling used vegetable oil from local restaurants is also an option for farmers running their equipment with biofuel.

Biopower is electricity produced from steam created by the burning of biomass or burning gas from biodigestors. Biodigestors trap and burn gasses from microbial decomposition of biomass such as manure. While the technology for biodigestors is fairly simple and has been used for many years, safe and clean systems have been designed for use today. Biomass crops such as switchgrass, corn, or fast growing trees can be pelletized and burned for heating buildings such as greenhouses or converted to electricity by way of steam. Some agricultural byproducts such as corn stovers can be used to produce energy without compromising soil quality or increasing erosion. Again, a cooperative purchase of a pelletizer may be sufficient for farmers wishing to burn pellets for heat. Electricity production from biomass for use by the public generally requires a large scale system owned by utility companies where farmers can sell their biomass products. Transportation costs in areas with smaller farms such as Massachusetts (see Figure 1) will hinder large scale electricity production from biomass, however, on farm production is still feasible.

GEOTHERMAL

While many of us think of hot springs and natural hot water as the source for geothermal power, there is a source of geothermal power much more commonly available which is simply the constant ground temperature. Farm buildings and homes can use geothermal heat pumps to exchange air temperature and ground temperature year round, keeping buildings cool in summer and warm in winter. While geothermal systems are more expensive to purchase and install than a typical fossil fuel burning furnace, the payback time is 5-10 years given the free fuel. A geothermal system is best suited for new construction considering the extensive excavation process involved.

Hydroelectric

Installing a hydroelectric system on a farm depends greatly on a good source of water with a constant flow. If these conditions exist and the farm's power needs are well documented, then the farmer may choose from several mechanisms: waterwheels, impulse turbines, or reaction turbines. While waterwheels are less efficient than turbines and are most suitable for powering grinders, they are relatively simple to use and can handle a wide range in water flow and debris. Turbines require a more controlled and constant rate of flow as well as a greater head, or elevation difference from the source of the water to the turbine blades. In general, the larger a hydroelectric system is, the cheaper per kWh. Small hydroelectric systems are a long term investment and 20-25 years is the general lifespan; so plan accordingly.

Solar

The sun's energy can be used for passive heating such as greenhouses, as solar thermal heating for hot water systems, or with photovoltaics (PV) it can be used to convert produce electricity. While photovoltaics are more expensive, technical advancements are making them a cheaper option. PV can be used to power lighting, electric fencing, small motors, fans, pumping water, or charging batteries. In rural areas or sections of the farm away from power lines PV can be the only option.

Passive solar design can be used to heat greenhouses; a cost effective option for small growers interested in extending their growing season. A passive solar greenhouse is highly insulated, uses natural ventilation, and has glass panels angled to collect maximum solar heat depending on the latitude or location of the greenhouse. Solar thermal systems for heating water work in a similar fashion to passive solar by collecting and storing heat. On a dairy farm where up to 40% of the energy used is for water heating, a solar water heater can reduce heating costs up to 85% annually. Like any major appliance, certified and labeled solar heaters are the most reliable. Consult the Solar Rating and Certification Corporation (SRCC)(link is external) for certification and rating of solar water heaters.

Wind

Turbines used to produce electricity from wind can provide a large portion of the average power needs of a farm, however must be located in high wind areas and generally require at least one acre of land to produce enough energy. Farmers or ranchers with grazing land in an area with good wind speeds could also consider leasing their land for wind power production to a utility company while still being able to use their land for their animals. For the option of running a system connected to the grid, the land owner must have an average annual wind speed of at least 10 mph. Installing wind turbines is a long term investment; costing from $13,000 to $40,000 for a residential wind energy system. However, the benefits of a wind turbine can outweigh the costs of installation if local utility costs are high, you chose a turbine that fits the wind speed of your location, and ultimately on how you use and maintain your system.

Keep in mind that a wind turbine can be a good compliment to a photovoltaic system in temperate climates. Wind energy is available when solar energy is not. It is strongest in Fall, Winter, and Spring, as well as at night when hot air rises from the earth's surface, increasing air flow.

2

AGRICULTURAL MACHINERY

Agricultural machinery relates to the mechanical structures and devices used in farming or other agriculture. There are many types of such equipment, from hand tools and power tools to tractors and the countless kinds of farm implements that they tow or operate. Diverse arrays of equipment are used in both organic and nonorganic farming. Especially since the advent of mechanised agriculture, agricultural machinery is an indispensable part of how the world is fed.

THE INDUSTRIAL REVOLUTION

With the coming of the Industrial Revolution and the development of more complicated machines, farming methods took a great leap forward. Instead of harvesting grain by hand with a sharp blade, wheeled machines cut a continuous swath. Instead of threshing the grain by beating it with sticks, threshing machines separated the seeds from the heads and stalks. The first tractors appeared in the late 19th century.

Steam power

Power for agricultural machinery was originally supplied by ox or other domesticated animals. With the invention of steam power came the portable engine, and later the traction engine, a multipurpose, mobile energy source that was the ground-crawling cousin to the steam locomotive. Agricultural steam engines took over the heavy pulling work of oxen, and were also equipped with a pulley that could power stationary machines via the use of a long belt. The steam-powered machines were low-powered by today's standards but because of their size and their low gear ratios, they could provide a large drawbar pull. The slow speed of steam-powered machines led farmers to comment that tractors had two speeds: "slow, and damn slow".

Internal combustion engines

The internal combustion engine; first the petrol engine, and later diesel engines; became the main source of power for the next generation of tractors. These engines also contributed to the development of the self-propelled combine harvester and thresher, or the combine harvester (also shortened to 'combine'). Instead of cutting the grain stalks and transporting them to a stationary threshing machine, these combines cut, threshed, and separated the grain while moving continuously throughout the field.

AGRICULTURAL MACHINERY TYPES

Tractors

Tractors do the majority of work on a modern farm. They are used to push/pull implements—machines that till the ground, plant seed, and perform other tasks. Tillage implements prepare the soil for planting by loosening the soil and killing weeds or competing plants. The best-known is the plow, the ancient implement that was upgraded in 1838 by John Deere. Plows are now used less frequently in the U.S. than formerly, with offset disks used instead to turn over the soil, and chisels used to gain the depth needed to retain moisture.

Combines

Combine is a machine designed to efficiently harvest a variety of grain crops. The name derives from its combining four separate harvesting operations—reaping, threshing, gathering, and winnowing—into a single process. Among the crops harvested with a combine are wheat, rice, oats, rye, barley, corn (maize), sorghum, soybeans, flax (linseed), sunflowers and rapeseed.

Planters

The most common type of seeder is called a planter, and spaces seeds out equally in long rows, which are usually two to three feet apart. Some crops are planted by drills, which put out much more seed in rows less than a foot apart, blanketing the field with crops. Transplanters automate the task of transplanting seedlings to the field. With the widespread use of plastic mulch, plastic mulch layers, transplanters, and seeders lay down long rows of plastic, and plant through them automatically.

Sprayers

After planting, other agricultural machinery such as self-propelled sprayers can be used to apply fertilizer and pesticides. Agriculture sprayer application is a method to protect crops from weeds by using herbicides, fungicides, and insecticides. Spraying or planting a cover crop are ways to mix weed growth.

Balers and other agriculture implements

Planting crop Hay balers can be used to tightly package grass or alfalfa into a storable form for the winter months. Modern irrigation relies on machinery. Engines, pumps and other specialized gear provide water quickly and in high volumes to large areas of land.

Similar types of equipment such as agriculture sprayers can be used to deliver fertilizers and pesticides.

Besides the tractor, other vehicles have been adapted for use in farming, including trucks, airplanes, and helicopters, such as for transporting crops and making equipment mobile, to aerial spraying and livestock herd management.

NEW TECHNOLOGY AND THE FUTURE

A New Holland TR85 combine harvester

The basic technology of agricultural machines has changed little in the last century. Though modern harvesters and planters may do a better job or be slightly tweaked from their predecessors, the US$250,000 combine of today still cuts, threshes, and separates grain in the same way it has always been done. However, technology is changing the way that humans operate the machines, as computer monitoring systems, GPS locators and self-steer programs allow the most advanced tractors and implements to be more precise and less wasteful in the use of fuel, seed, or fertilizer. In the foreseeable future, there may be mass production of driverless tractors, which use GPS maps and electronic sensors.

Open source agricultural equipment

Many farmers are upset by their inability to fix the new types of high-tech farm equipment. This is due mostly to companies using intellectual property law to prevent farmers from having the legal right to fix their equipment (or gain access to the information to allow them to do it). In October 2015 an exemption was added to the DMCA to allow inspection and modification of the software in cars and other vehicles including agricultural machinery.

The Open Source Agriculture movement counts different initiatives and organizations such as Farm Labs which is a network in Europe, l'Atelier Paysan which is a cooperative to teach farmers in France how to build and repair their tools, and Ekylibre which is an open-source company to provide farmers in France with open source software (SaaS) to manage farming operations. In the United States, the MIT Media Lab's Open Agriculture Initiative seeks to foster "the creation of an open-source ecosystem of technologies that enable and promote transparency, networked experimentation, education, and hyper-local production". It develops the Personal Food Computer, an educational project to create a "controlled environment agriculture technology platform that uses robotic systems to control and monitor climate, energy, and plant growth inside of a specialized growing chamber". It includes the development of Open Phenom, an open source library with open data sets for climate recipes which link the phenotype response of plants (taste, nutrition) to environmental variables, biological, genetic and resource-related necessary for cultivation (input). Plants with the same genetics can naturally vary in color, size, texture, growth rate, yield, flavor, and nutrient density according to the environmental conditions in which they are produced.

AGRICULTURAL MACHINERY INDUSTRY

The agricultural machinery industry or agricultural engineering industry is the part of the industry, that produces and maintain tractors, agricultural machinery and agricultural implements used in farming or other agriculture. This branch is considered to be part of the machinery industry.

19th century

The agricultural machinery industry emerged in Britain and the United States in the 19th century. Until then the common tools of farming were the plough and the sickle. These iron agricultural implements were often made by blacksmiths in the local village, who regularly also acted as farrier. In the first part of the 19th century some of the early agricultural machine manufacturers arose from these blacksmith workshops, such as John Deere who started up with the production of ploughs in series in the 1840s.

Agricultural Implements, 1851.

Other companies arose from the introduction of horse drawn reaping, which replaced the type of hand reaper in use since biblical times. A company as the McCormick started up with building these kind of harvesting machines around the 1840s. And another origin of agricultural industry was the introduction of combined harvesting, threshing and cleaning in the 1830s. The Case Corporation for example started building those in 1842 as the Racine Threshing Machine Works. Until early 20th century most of those machinery were powered by horses.

Mid 19th century the portable steam-powered plowing engines were introduced. They were used in pairs, placed on either side of a field to haul a plow back and forth between them using a wire cable. These portable engines were also used to power threshing machines, mills and pumps. The portable steam engines were produced by specific agricultural machinery maker, such as Ransomes, Sims & Jefferies who had started as brass and iron-founder making casting ploughshares late 18th century.

Late 19th century in Britain more companies such as Richard Garrett & Sons and Mann's Patent Steam Cart and Wagon Company developed steam tractors for direct ploughing, but the heavy, wet soil of England meant that these designs were less economical than a team of horses. In the United States, where soil conditions permitted, steam tractors were used to direct-haul plows. Steam-powered agricultural engines remained in use well into the 20th century until reliable internal combustion engines had been developed.

Collins (1987) recalled that the impact of the agricultural machinery industry in the 19th century was still limited. He stated : "prior the third quarter of the nineteenth century the impact of machinery in agriculture was slight compared with that in manufacturing industry. Some operations such as barn work and hay and corn harvesting had been largely mechanized by 1880 but, up to the Second World War, many were still performed by hand labour and large numbers of workers were still required for seasonal tasks such as hop- and fruit-picking and vegetable cultivations."

20th century

In the beginning of the 20th century in the UK the Agricultural machinery industry "although composed of many hundreds of firms, was dominated by a few large ones, chiefly in the eastern counties of England. The total output of the industry was estimated to be worth 6.5 million pounds in 1913, or about 5 percent of the total value of the output of the mechanical engineering industry at the first Census of Production in 1907."

AGRICULTURAL MACHINERY MANUFACTURERS

Agricultural machinery manufacturers exist in sizes from small and medium business to multinationals. James & Akrasanee (1988) stipulated that those forms have different production management, and can be classified into three groups:

The first group consists of those workshops with limited and simple equipment. Despite their flexibility, the production management system is very rare in this case.

The second group includes those with certain degrees of division of labour within the plant. They do organize some plant layout, but not in a fully systematic manner.

The last group constitutes those with clear production lines and division of labour.

The two largest agricultural machinery manufacturers worldwide based on revenue (in 2012):

John Deere : $37.79 billion revenues and employing 67,000 people worldwide in 2013. Its workforce has increased significantly in the past decennia from 43.000 in 2002 to 67.000 in 2013.

CNH Industrial : $19.4 billion revenues and employing 33.800 people worldwide in 2013.

In the United States the revenue of tractor & agricultural machinery manufacturing sector increased from about 35 billion U.S. dollars in 2009 to 42 billion U.S. dollars in 2014. Global agricultural machinery industry

According to the CEMA, the European association representing the European agricultural machinery industry, Europe was the largest producer of agricultural machinery globally in 2013:

- Europe produced 31% of the global production
- North America produced 26.5% of the global production,
- China produced 26.5% of the global production, and

- The remaining 16% is produced elsewhere.

In total, the production of agricultural machinery worldwide in 2013 generated about 95 billion euros, based on a report Agricultural Machinery November 2013 : Market Perspectives 2014 by the VDMA, a German engineering association. Another recent statistic mentioned, that the agricultural machinery industry in 2010 generated worldwide about $56 billion.

AGRICULTURAL MACHINERY INDUSTRY IN EUROPE

With a volume of 28 billion euros (2014) agricultural machinery production in the European Union (EU) represented 28 per cent of world total according to VDMA. Out of the EU total, the share of Germany is 27 per cent, followed by Italy (17 per cent) and France (14 per cent). All leading international manufacturers have several production sites on the European continent, and their products from these factories in general are destined for the high end customers who yield for efficiency and accuracy on the fields.

The EU production of tractors was 187,000 units, correspondent to a value of 9.4 billion euros (2014). Outside the EU, a major production site for tractors is the Minsk Tractor Works (Belarus) with a high market share in the neighbouring countries, especially within the Commonwealth of Independent States. The leading countries for tractor production in the EU are Germany and Italy (with John Deere, Case New Holland, AGCO and Same Deutz-Fahr being the major brands). Germany is also the major place for the manufacturing of harvesting, spraying, fertilizing and seeding equipment. Italy and France have a primary position for soil working machines and transportation vehicles. The majority of the agricultural machinery companies are small and medium-sized and family owned.

One third of the EU production is destined to markets outside the union, mainly to the Eastern European markets, but also to North America and, increasingly, to Asia.

After a peak in 2003 and 29 billion euros of turnover, the sector experienced an economic downturn in 2014/2015. The strategy of the European industry is to keep its leading technological position and to diversify its market presence into the high potential regions (e.g. Asia).

Agricultural machinery industry in North America

In 2014, the AEM stated that "U.S. domestic sales of agricultural machinery and equipment rose from about $20 billion in 1999 to $38 billion in 2012. Since 1998 and through 2012, U.S. exports of agricultural equipment have grown relatively quickly, more than doubling from about $4 billion to $8.7 billion.

In 2021, the AEM reported an increase in sales of new agricultural machinery for the U.S. market. There also an 27% increase of new machinery manufacturing for this year. AEM members are also anticipating a 6% to 10% increase in agriculture machinery demand over the next 12 months. In the report is mentioned that in 2021 the U.S. is the world's leader in producing agricultural machinery.

Agricultural machinery industry in South America

Graciela summarized (2008) that "the agricultural machinery industry in Argentina depends for growth on higher exports and further progress towards internationalization, which are strategic goals for the largest firms. Given the dynamism of global demand for this type of machinery, the conclusion is that the sector can increase its sales in export markets, where some of its products are competing well."

Agricultural machinery industry in Asia

Yuan'en (2007) stated, that the "Chinese government has attached great importance to agriculture, rural areas and farmers, made a series of policies that favorite to the farmers, added the financial subsidy to the agriculture, especially constituted the subsidy policies for the farmers on their agricultural machinery purchasing."

A 2013-2014 report acknowledged that "China's agricultural machinery industry has achieved rapid development, the gross output value of agricultural machinery exceeded RMB 300 billion, total power of agricultural machinery surpassed 1 billion kilowatts, and the integrated mechanization level of agricultural crops exceeded 50%, demonstrating that China has entered into a mechanized production-oriented period.... Tractors and harvesting machinery are the most important agricultural machinery products in China, with annual output of 2.25 million units and 1.114 million units in 2012 separately...

Gyanendra (2006) recalled, that "the early agricultural mechanization in India was greatly influenced by the technological development in England. Irrigation pumps, tillage equipment, chaff cutters, tractors and threshers were gradually introduced for farm mechanization. The high yielding varieties with assured irrigation and higher rate of application of fertilizers gave higher returns that enabled farmers to adopt mechanization inputs, especially after Green revolution in 1960s." The Turkish Ministry of Economy (2011) explained that "Turkey is one of the few countries of the world which is self-sufficient in food. At present Turkey is the largest producer and exporter of agricultural products in the Near East and North Africa... The agricultural machinery industry is comprised of about 1000 manufacturers countrywide.... Companies manufacturing agricultural machinery and equipment except tractors are mostly small and medium size companies. The sector employs about 15,000 workers... The value of agricultural machinery and spare parts and agricultural tractors and trailers exports was US$ 313 million in 2009."

MECHANISED AGRICULTURE

Mechanised agriculture is the process of using agricultural machinery to mechanise the work of agriculture, greatly increasing farm worker productivity. In modern times, powered machinery has replaced many farm jobs formerly carried out by manual labour or by working animals such as oxen, horses and mules.

The entire history of agriculture contains many examples of the use of tools, such as the hoe and the plough. The ongoing integration of machines since the Industrial Revolution however has allowed farming to become much less labour-intensive.

Current mechanised agriculture includes the use of tractors, trucks, combine harvesters, countless types of farm implements, aeroplanes and helicopters (for aerial application), and other vehicles. Precision agriculture even uses computers in conjunction with satellite imagery and satellite navigation (GPS guidance) to increase yields.

Mechanisation was one of the large factors responsible for urbanisation and industrial economies. Besides improving production efficiency, mechanisation encourages large scale production and sometimes can improve the quality of farm produce. On the other hand, it can displace unskilled farm labour and can cause environmental degradation (such as pollution, deforestation, and soil erosion), especially if it is applied shortsightedly rather than holistically.

Jethro Tull's seed drill (ca. 1701) was a mechanical seed spacing and depth placing device that increased crop yields and saved seed. It was an important factor in the British Agricultural Revolution.

Since the beginning of agriculture threshing was done by hand with a flail, requiring a great deal of labour. The threshing machine, which was invented in 1794 but not widely used for several more decades, simplified the operation and allowed the use of animal power. Before the invention of the grain cradle (ca. 1790) an able bodied labourer could reap about one quarter acre of wheat in a day using a sickle. It was estimated that each of Cyrus McCormick's horse-pulled reapers (ca. 1830s) freed up five men for military service in the US Civil War. Later innovations included raking and binding machines. By 1890 two men and two horses could cut, rake and bind 20 acres of wheat per day.

In the 1880s the reaper and threshing machine were combined into the combine harvester. These machines required large teams of horses or mules to pull. Steam power was applied to threshing machines in the late 19th century. There were steam engines that moved around on wheels under their own power for supplying temporary power to stationary threshing machines. These were called road engines, and Henry Ford seeing one as a boy was inspired to build an automobile.

With internal combustion came the first modern tractors in the early 1900s, becoming more popular after the Fordson tractor (ca. 1917). At first reapers and combine harvesters were pulled by teams of horses or tractors, but in the 1930s self powered combines were developed.

Advertising for motorised equipment in farm journals during this era did its best to compete against horse-drawn methods with economic arguments, extolling common themes such as that a tractor "eats only when it works", that one tractor could replace many horses, and that mechanisation could allow one man to get more work done per day than he ever had before. The horse population in the US began to decline in the 1920s after the conversion of agriculture and transportation to internal combustion. Peak tractor sales in the US were around 1950. In addition to saving labour, this freed up much land previously used for supporting draft animals. The greatest period of growth in agricultural productivity in the US was from the 1940s to the 1970s, during which time agriculture was benefiting from internal combustion powered tractors and combine harvesters, chemical fertilisers and the green revolution.

Although farmers of corn, wheat, soy, and other commodity crops had replaced most of their workers with harvesting machines and combines by the 1950s enabling them to efficiently cut and gather grains, growers of produce continued to rely on human pickers to avoid the bruising of the product in order to maintain the blemish-free appearance demanded by customers. The continuous supply of illegal workers from Latin America that were willing to harvest the crops for low wages further suppressed the need for mechanisation. As the number of illegal workers has continued to decline since reaching its peak in 2007 due to increased border patrols and an improving Mexican economy, the industry is increasing the use of mechanisation. Proponents argue that mechanisation will boost productivity and help to maintain low food prices while farm worker advocates assert that it will eliminate jobs and will give an advantage to large growers who are able to afford the required equipment.

Harvesting

Asparagus are presently harvested by hand with labour costs at 71% of production costs and 44% of selling costs. Asparagus is a difficult crop to harvest since each spear matures at a different speed making it difficult to achieve a uniform harvest. A prototype asparagus harvesting machine – using a light-beam sensor to identify the taller spears – is expected to be available for commercial use.

Mechanization of Maine's blueberry industry has reduced the number of migrant workers required from 5,000 in 2005 to 1,500 in 2015 even though production has increased from 50-60 million pounds per year in 2005 to 90 million pounds in 2015.

As of 2014, prototype chili pepper harvesters are being tested by New Mexico State University. The New Mexico green chile crop is currently hand-picked entirely by field workers as chili pods tend to bruise easily. The first commercial application commenced in 2015. The equipment is expected to increase yield per acre and help to offset a sharp decline in acreage planted due to the lack of available labour and drought conditions.

As of 2010, approximately 10% of the processing orange acreage in Florida is harvested mechanically, mainly with citrus canopy shaker machines. Mechanization has progressed slowly due to the uncertainty of future economic benefits due to competition from Brazil and the transitory damage to orange trees when they are harvested.

There has been an ongoing transition to mechanical harvesting of cling peaches (mostly used in canning) where the cost of labor is 70 percent of a grower's direct costs. In 2016, 12 percent of the cling peach tonnage from Yuba County and Sutter County in California will be mechanically harvested. Fresh peaches destined for direct customer sales must still be hand-picked.

As of 2007, mechanised harvesting of raisins is at 45%; however the rate has slowed due to high raisin demand and prices making the conversion away from hand labour less urgent. A new strain of grape developed by the USDA that drys on the vine and is easily harvested mechanically is expected to reduce the demand for labour.

Strawberries are a high cost-high value crop with the economics supporting mechanisation. In 2005, picking and hauling costs were estimated at $594 per ton or 51% of the total grower cost. However, the delicate nature of fruit make it an unlikely candidate for mechanisation in the near future. A strawberry harvester developed by Shibuya Seiki and unveiled in Japan in 2013 is able to pick a strawberry every eight seconds. The robot identifies which strawberries are ready to pick by using three separate cameras and then once identified as ready, a mechanised arm snips the fruit free and gently places it in a basket. The robot moves on rails between the rows of strawberries which are generally contained within elevated greenhouses. The machine costs 5 million yen. A new strawberry harvester made by Agrobot that will harvest strawberries on raised, hydroponic beds using 60 robotic arms is expected to be released in 2016.

Mechanical harvesting of tomatoes started in 1965 and as of 2010, nearly all processing tomatoes are mechanically harvested. As of 2010, 95% of the US processed tomato crop is produced in California. Although fresh market tomatoes have substantial hand harvesting costs (in 2007, the costs of hand picking and hauling were $86 per ton which is 19% of total grower cost), packing and selling costs were more of a concern (at 44% of total grower cost) making it likely that cost saving efforts would be applied there.

According to a 1977 report by the California Agrarian Action Project, during the summer of 1976 in California, many harvest machines had been equipped with a photo-electric scanner that sorted out green tomatoes among the ripe red ones using infrared lights and colour sensors. It worked in lieu of 5,000 hand harvesters causing displacement of innumerable farm labourers as well as wage cuts and shorter work periods. Migrant workers were hit the hardest. To withstand the rigour of the machines, new crop varieties were bred to match the automated pickers. UC Davis Professor G.C. Hanna propagated a thick-skinned tomato called VF-145. But even still, millions were damaged with impact cracks and university breeders produced a tougher and juiceless "square round" tomato. Small farms were of insufficient size to obtain financing to purchase the equipment and within 10 years, 85% of the state's 4,000 cannery tomato farmers were out of the business. This led to a concentrated tomato industry in California that "now packed 85% of the nation's tomato products". The monoculture fields fostered rapid pest growth, requiring the use of "more than four million pounds of pesticides each year" which greatly affected the health of the soil, the farm workers, and possibly the customers.

PRODUCTIVITY-IMPROVING TECHNOLOGIES

Productivity is often measured as the ratio of (aggregate) output to (aggregate) input in the production of goods and services. Productivity is increased by lowering the amount of labor, capital, energy or materials that go into producing any given amount of economic goods and services. Increases in productivity are largely responsible for the increase in per capita living standards.

Productivity-improving technologies date back to antiquity, with rather slow progress until the late Middle Ages. Important examples of early to medieval European technology include the water wheel, the horse collar, the spinning wheel, the three-field system (after 1500 the four-field system—see crop rotation) and the blast furnace. All of these technologies had been in use in China, some for centuries, before being introduced to Europe.

Technological progress was aided by literacy and the diffusion of knowledge that accelerated after the spinning wheel spread to Western Europe in the 13th century. The spinning wheel increased the supply of rags used for pulp in paper making, whose technology reached Sicily sometime in the 12th century. Cheap paper was a factor in the development of the movable type printing press, which led to a large increase in the number of books and titles published. Books on science and technology eventually began to appear, such as the mining technical manual De Re Metallica, which was the most important technology book of the 16th century and was the standard chemistry text for the next 180 years.

Francis Bacon (1561–1626) is known for the scientific method, which was a key factor in the scientific revolution. Bacon stated that the technologies that distinguished Europe of his day from the Middle Ages were paper and printing, gunpowder and the magnetic compass, known as the four great inventions. The four great inventions important to the development of Europe were of Chinese origin. Other Chinese inventions included the horse collar, cast iron, an improved plow and the seed drill. See also List of Chinese inventions.

Mining and metal refining technologies played a key role in technological progress. Much of our understanding of fundamental chemistry evolved from ore smelting and refining, with De Re Metallica being the leading chemistry text. Railroads evolved from mine carts and the first steam engines were designed specifically for pumping water from mines. The significance of the blast furnace goes far beyond its capacity for large scale production of cast iron. The blast furnace was the first example of continuous production and is a countercurrent exchange process, various types of which are also used today in chemical and petroleum refining. Hot blast, which recycled what would have otherwise been waste heat, was one of engineering's key technologies. It had the immediate effect of dramatically reducing the energy required to produce pig iron, but reuse of heat was eventually applied to a variety of industries, particularly steam boilers, chemicals, petroleum refining and pulp and paper.

Before the 17th century scientific knowledge tended to stay within the intellectual community, but by this time it became accessible to the public in what is called "open science". Near the beginning of the Industrial Revolution came publication of the Encyclopédie, written by numerous contributors and edited by Denis Diderot and Jean le Rond d'Alembert (1751–72). It contained many articles on science and was the first general encyclopedia to provide in depth coverage on the mechanical arts, but is far more recognized for its presentation of thoughts of the Enlightenment.

Economic historians generally agree that, with certain exceptions such as the steam engine, there is no strong linkage between the 17th century scientific revolution (Descartes, Newton, etc.) and the Industrial Revolution. However, an important mechanism for the transfer of technical knowledge was scientific societies, such as The Royal Society of London for Improving Natural Knowledge, better known as the Royal Society, and the Académie des Sciences. There were also technical colleges, such as the École Polytechnique. Scotland was the first place where science was taught (in the 18th century) and was where Joseph Black discovered heat capacity and latent heat and where his friend James Watt used knowledge of heat to conceive the separate condenser as a means to improve the efficiency of the steam engine.

Probably the first period in history in which economic progress was observable after one generation was during the British Agricultural Revolution in the 18th century. However, technological and economic progress did not proceed at a significant rate until the English Industrial Revolution in the late 18th century, and even then productivity grew about 0.5% annually. High productivity growth began during the late 19th century in what is sometimes called the Second Industrial Revolution. Most major innovations of the Second Industrial Revolution were based on the modern scientific understanding of chemistry, electromagnetic theory and thermodynamics and other principles known to the profession of engineering.

New forms of energy and power

Before the industrial revolution the only sources of power were water, wind and muscle. Most good water power sites (those not requiring massive modern dams) in Europe were developed during the medieval period. In the 1750s John Smeaton, the "father of civil engineering," significantly improved the efficiency of the water wheel by applying scientific principles, thereby adding badly needed power for the Industrial Revolution. However water wheels remained costly, relatively inefficient and not well suited to very large power dams. Benoît Fourneyron's highly efficient turbine developed in the late 1820s eventually replaced waterwheels. Fourneyron type turbines can operate at 95% efficiency and used in today's large hydro-power installations. Hydro-power continued to be the leading source of industrial power in the United States until past the mid 19th century because of abundant sites, but steam power overtook water power in the UK decades earlier.

In 1711 a Newcomen steam engine was installed for pumping water from a mine, a job that typically was done by large teams of horses, of which some mines used as many as 500. Animals convert feed to work at an efficiency of about 5%, but while this was much more than the less than 1% efficiency of the early Newcomen engine, in coal mines there was low quality coal with little market value available. Fossil fuel energy first exceeded all animal and water power in 1870. The role energy and machines replacing physical work is discussed in Ayres-Warr (2004, 2009).

While steamboats were used in some areas, as recently as the late 19th Century thousands of workers pulled barges. Until the late 19th century most coal and other minerals were mined with picks and shovels and crops were harvested and grain threshed using animal power or by hand. Heavy loads like 382 pound bales of cotton were handled on hand trucks until the early 20th century.

A young "drawer" pulling a coal tub along a mine gallery. Minecarts were more common than the skid shown. Railroads descended from minecarts. In Britain laws passed in 1842 and 1844 improved working conditions in mines.

Excavation was done with shovels until the late 19th century when steam shovels came into use. It was reported that a laborer on the western division of the Erie Canal was expected to dig 5 cubic yards per day in 1860; however, by 1890 only 3-1/2 yards per day were expected. Today's large electric shovels have buckets that can hold 168 cubic meters (220 cubic yards) and consume the power of a city of 100,000. Dynamite, a safe to handle blend of nitroglycerin and diatomaceous earth was patented in 1867 by Alfred Nobel. Dynamite increased productivity of mining, tunneling, road building, construction and demolition and made projects such as the Panama Canal possible.

Steam power was applied to threshing machines in the late 19th century. There were steam engines that moved around on wheels under their own power that were used for supplying temporary power to stationary farm equipment such as threshing machines. These were called road engines, and Henry Ford seeing one as a boy was inspired to build an automobile. Steam tractors were used but never became popular.

With internal combustion came the first mass-produced tractors (Fordson c. 1917). Tractors replaced horses and mules for pulling reapers and combine harvesters, but in the 1930s self powered combines were developed. Output per man hour in growing wheat rose by a factor of about 10 from the end of World War II until about 1985, largely because of powered machinery, but also because of increased crop yields. Corn manpower showed a similar but higher productivity increase. See below:Mechanized agriculture.

One of the greatest periods of productivity growth coincided with the electrification of factories which took place between 1900 and 1930 in the U.S.

ENERGY EFFICIENCY

In engineering and economic history the most important types of energy efficiency were in the conversion of heat to work, the reuse of heat and the reduction of friction. There was also a dramatic reduction energy required to transmit electronic signals, both voice and data.

Conversion of heat to work

The early Newcomen steam engine was about 0.5% efficient and was improved to slightly over 1% by John Smeaton before Watt's improvements, which increased thermal efficiency to 2%. In 1900 it took 7 lbs coal/ kw hr.

Electrical generation was the sector with the highest productivity growth in the U.S. in the early twentieth century. After the turn of the century large central stations with high pressure boilers and efficient steam turbines replaced reciprocating steam engines and by 1960 it took 0.9 lb coal per kw-hr. Counting the improvements in mining and transportation the total improvement was by a factor greater than 10. Today's steam turbines have efficiencies in the 40% range. Most electricity today is produced by thermal power stations using steam turbines.

The Newcomen and Watt engines operated near atmospheric pressure and used atmospheric pressure, in the form of a vacuum caused by condensing steam, to do work. Higher pressure engines were light enough, and efficient enough to be used for powering ships and locomotives. Multiple expansion (multi-stage) engines were developed in the 1870s and were efficient enough for the first time to allow ships to carry more freight than coal, leading to great increases in international trade.

The first important diesel ship was the MS Selandia launched in 1912. By 1950 one-third of merchant shipping was diesel powered. Today the most efficient prime mover is the two stroke marine diesel engine developed in the 1920s, now ranging in size to over 100,000 horsepower with a thermal efficiency of 50%.

Steam locomotives that used up to 20% of the U.S. coal production were replaced by diesel locomotives after World War II, saving a great deal of energy and reducing manpower for handling coal, boiler water and mechanical maintenance.

Improvements in steam engine efficiency caused a large increase in the number of steam engines and the amount of coal used, as noted by William Stanley Jevons in The Coal Question. This is called the Jevons paradox.

ELECTRIFICATION AND THE PRE-ELECTRIC TRANSMISSION OF POWER

Further information: Line shaft

Electricity consumption and economic growth are strongly correlated. Per capita electric consumption correlates almost perfectly with economic development. Electrification was the first technology to enable long-distance transmission of power with minimal power losses. Electric motors did away with line shafts for distributing power and dramatically increased the productivity of factories. Very large central power stations created economies of scale and were much more efficient at producing power than reciprocating steam engines. Electric motors greatly reduced the capital cost of power compared to steam engines.

The main forms of pre-electric power transmission were line shafts, hydraulic power networks and pneumatic and wire rope systems. Line shafts were the common form of power transmission in factories from the earliest industrial steam engines until factory electrification. Line shafts limited factory arrangement and suffered from high power losses. Hydraulic power came into use in the mid 19th century. It was used extensively in the Bessemer process and for cranes at ports, especially in the UK. London and a few other cities had hydraulic utilities that provided pressurized water for industrial over a wide area.

Pneumatic power began being used industry and in mining and tunneling in the last quarter of the 19th century. Common applications included rock drills and jack hammers. Wire ropes supported by large grooved wheels were able to transmit power with low loss for a distance of a few miles or kilometers. Wire rope systems appeared shortly before electrification.

Reuse of heat

Recovery of heat for industrial processes was first widely used as hot blast in blast furnaces to make pig iron in 1828. Later heat reuse included the Siemens-Martin process which was first used for making glass and later for steel with the open hearth furnace. (See: Iron and steel below). Today heat is reused in many basic industries such as chemicals, oil refining and pulp and paper, using a variety of methods such as heat exchangers in many processes. Multiple-effect evaporators use vapor from a high temperature effect to evaporate a lower temperature boiling fluid. In the recovery of kraft pulping chemicals the spent black liquor can be evaporated five or six times by reusing the vapor from one effect to boil the liquor in the preceding effect. Cogeneration is a process that uses high pressure steam to generate electricity and then uses the resulting low pressure steam for process or building heat.

Industrial process have undergone numerous minor improvements which collectively made significant reductions in energy consumption per unit of production.

Reducing friction

Reducing friction was one of the major reasons for the success of railroads compared to wagons. This was demonstrated on an iron plate covered wooden tramway in 1805 at Croydon, U.K.

" A good horse on an ordinary turnpike road can draw two thousand pounds, or one ton. A party of gentlemen were invited to witness the experiment, that the superiority of the new road might be established by ocular demonstration. Twelve wagons were loaded with stones, till each wagon weighed three tons, and the wagons were fastened together. A horse was then attached, which drew the wagons with ease, six miles in two hours, having stopped four times, in order to show he had the power of starting, as well as drawing his great load."

Better lubrication, such as from petroleum oils, reduced friction losses in mills and factories. Anti-friction bearings were developed using alloy steels and precision machining techniques available in the last quarter of the 19th century. Anti-friction bearings were widely used on bicycles by the 1880s. Bearings began being used on line shafts in the decades before factory electrification and it was the pre-bearing shafts that were largely responsible for their high power losses, which were commonly 25 to 30% and often as much as 50%.

Lighting efficiency

Electric lights were far more efficient than oil or gas lighting and did not generate smoke, fumes nor as much heat. Electric light extended the work day, making factories, businesses and homes more productive. Electric light was not a great fire hazard like oil and gas light.

The efficiency of electric lights has continuously improved from the first incandescent lamps to tungsten filament lights. The fluorescent lamp, which became commercial in the late 1930s, is much more efficient than incandescent lighting. Light-emitting diodes or LED's are highly efficient and long lasting.

Infrastructures

The relative energy required for transport of a tonne-km for various modes of transport are: pipelines=1(basis), water 2, rail 3, road 10, air 100.

Roads

Unimproved roads were extremely slow, costly for transport and dangerous. In the 18th century layered gravel began being increasingly used, with the three layer Macadam coming into use in the early 19th century. These roads were crowned to shed water and had drainage ditches along the sides. The top layer of stones eventually crushed to fines and smoothed the surface somewhat. The lower layers were of small stones that allowed good drainage. Importantly, they offered less resistance to wagon wheels and horses hooves and feet did not sink in the mud. Plank roads also came into use in the U.S. in the 1810s-1820s. Improved roads were costly, and although they cut the cost of land transportation in half or more, they were soon overtaken by railroads as the major transportation infrastructure.

MECHANIZATION

The seed drill is a mechanical device for spacing and planting seed at the appropriate depth. It originated in ancient China before the 1st century BC. Saving seed was extremely important at a time when yields were measured in terms of seeds harvested per seed planted, which was typically between 3 and 5. The seed drill also saved planting labor. Most importantly, the seed drill meant crops were grown in rows, which reduced competition of plants and increase yields. It was reinvented in 16th century Europe based on verbal descriptions and crude drawings brought back from China. Jethro Tull patented a version in 1700; however, it was expensive and unreliable. Reliable seed drills appeared in the mid 19th century.

Since the beginning of agriculture threshing was done by hand with a flail, requiring a great deal of labor. The threshing machine (ca. 1794) simplified the operation and allowed it to use animal power. By the 1860s threshing machines were widely introduced and ultimately displaced as much as a quarter of agricultural labor. In Europe, many of the displaced workers were driven to the brink of starvation.

Threshing machine from 1881. Steam engines were also used instead of horses. Today both threshing and reaping are done with a combine harvester.

Before c. 1790 a worker could harvest 1/4 acre per day with a scythe. In the early 1800s the grain cradle was introduced, significantly increasing the productivity of hand labor. It was estimated that each of Cyrus McCormick's horse pulled reapers (Ptd. 1834) freed up five men for military service in the U.S. Civil War. By 1890 two men and two horses could cut, rake and bind 20 acres of wheat per day. In the 1880s the reaper and threshing machine were combined into the combine harvester. These machines required large teams of horses or mules to pull. Over the entire 19th century the output per man hour for producing wheat rose by about 500% and for corn about 250%.

Harvesting oats in a Claas Lexion 570 combine with enclosed, air-conditioned cab with rotary thresher and laser-guided hydraulic steering

Farm machinery and higher crop yields reduced the labor to produce 100 bushels of corn from 35 to 40 hours in 1900 to 2 hours 45 minutes in 1999. The conversion of agricultural mechanization to internal combustion power began after 1915. The horse population began to decline in the 1920s after the conversion of agriculture and transportation to internal combustion. In addition to saving labor, this freed up much land previously used for supporting draft animals.

The peak years for tractor sales in the U.S. were the 1950s. There was a large surge in horsepower of farm machinery in the 1950s.

Industrial machinery

The most important mechanical devices before the Industrial Revolution were water and wind mills. Water wheels date to Roman times and windmills somewhat later. Water and wind power were first used for grinding grain into flour, but were later adapted to power trip hammers for pounding rags into pulp for making paper and for crushing ore. Just before the Industrial revolution water power was applied to bellows for iron smelting in Europe. (Water powered blast bellows were used in ancient China.) Wind and water power were also used in sawmills. The technology of building mills and mechanical clocks was important to the development of the machines of the Industrial Revolution.

The spinning wheel was a medieval invention that increased thread making productivity by a factor greater than ten. One of the early developments that preceded the Industrial Revolution was the stocking frame (loom) of c. 1589. Later in the Industrial Revolution came the flying shuttle, a simple device that doubled the productivity of weaving. Spinning thread had been a limiting factor in cloth making requiring 10 spinners using the spinning wheel to supply one weaver. With the spinning jenny a spinner could spin eight threads at once. The water frame (Ptd. 1768) adapted water power to spinning, but it could only spin one thread at a time. The water frame was easy to operate and many could be located in a single building. The spinning mule (1779) allowed a large number of threads to be spun by a single machine using water power. A change in consumer preference for cotton at the time of increased cloth production resulted in the invention of the cotton gin (Ptd. 1794). Steam power eventually was used as a supplement to water during the Industrial Revolution, and both were used until electrification. A graph of productivity of spinning technologies can be found in Ayres (1989), along with much other data related this article.

With a cotton gin (1792) in one day a man could remove seed from as much upland cotton as would have previously taken a woman working two months to process at one pound per day using a roller gin.

- An early example of a large productivity increase by special purpose machines is the c. 1803 Portsmouth Block Mills. With these machines 10 men could produce as many blocks as 110 skilled craftsmen.
- In the 1830s several technologies came together to allow an important shift in wooden building construction. The circular saw (1777), cut nail machines (1794), and steam engine allowed slender pieces of lumber such as 2"x4"s to be efficiently produced and then nailed together in what became known as balloon framing

(1832). This was the beginning of the decline of the ancient method of timber frame construction with wooden joinery.

- Following mechanization in the textile industry was mechanization of the shoe industry.
- The sewing machine, invented and improved during the early 19th century and produced in large numbers by the 1870s, increased productivity by more than 500%. The sewing machine was an important productivity tool for mechanized shoe production.
- With the widespread availability of machine tools, improved steam engines and inexpensive transportation provided by railroads, the machinery industry became the largest sector (by profit added) of the U. S. economy by the last quarter of the 19th century, leading to an industrial economy.

MACHINE TOOLS

Machine tools, which cut, grind and shape metal parts, were another important mechanical innovation of the Industrial Revolution. Before machine tools it was prohibitively expensive to make precision parts, an essential requirement for many machines and interchangeable parts. Historically important machine tools are the screw-cutting lathe, milling machine and metal planer (metalworking), which all came into use between 1800 and 1840. However, around 1900, it was the combination of small electric motors, specialty steels and new cutting and grinding materials that allowed machine tools to mass-produce steel parts. Production of the Ford Model T required 32,000 machine tools.

Modern manufacturing began around 1900 when machines, aided by electric, hydraulic and pneumatic power, began to replace hand methods in industry. An early example is the Owens automatic glass bottle blowing machine, which reduced labor in making bottles by over 80%. See also: Mass production#Factory electrification

Mining

Large mining machines, such as steam shovels, appeared in the mid-nineteenth century, but were restricted to rails until the widespread introduction of continuous track and pneumatic tires in the late 19th and early 20th centuries. Until then much mining work was mostly done with pneumatic drills, jackhammers, picks and shovels.

Coal seam undercutting machines appeared around 1890 and were used for 75% of coal production by 1934. Coal loading was still being done manually with shovels around 1930, but mechanical pick up and loading machines were coming into use. The use of the coal boring machine improved productivity of sub-surface coal mining by a factor of three between 1949 and 1969.

There is currently a transition going under way from more labor-intensive methods of mining to more mechanization and even automated mining.

Mechanized Materials Handling

Bulk materials handling

Unloading cotton c. 1900. Hydraulic cranes were in use in the U.K. for loading ships by the 1840s, but were little used in the U.S. Steam powered conveyors and cranes were used in the U.S. by the 1880s. In the early 20th century, electric operated cranes and motorized mobile loaders such as forklifts were used. Today non-bulk freight is containerized.

- Dry bulk materials handling systems use a variety of stationary equipment such as conveyors, stackers, reclaimers and mobile equipment such as power shovels and loaders to handle high volumes of ores, coal, grains, sand, gravel, crushed stone, etc. Bulk materials handling systems are used at mines, for loading and unloading ships and at factories that process bulk materials into finished goods, such as steel and paper mills.
- Mechanical stokers for feeding coal to locomotives were in use in the 1920s. A completely mechanized and automated coal handling and stoking system was first used to feed pulverized coal to an electric utility boiler in 1921. Liquids and gases are handled with centrifugal pumps and compressors, respectively.
- Conversion to powered material handling increased during WW 1 as shortages of unskilled labor developed and unskilled wages rose relative to skilled labor.
- A noteworthy use of conveyors was Oliver Evans's automatic flour mill built in 1785.
- Around 1900 various types of conveyors (belt, slat, bucket, screw or auger), overhead cranes and industrial trucks began being used for handling materials and goods in various stages of production in factories. See: Types of conveyor systems. and mass production.
- A well known application of conveyors is Ford. Motor Co.'s assembly line (c. 1913), although Ford used various industrial trucks, overhead cranes, slides and whatever devices necessary to minimize labor in handling parts in various parts of the factory.

Cranes

Cranes are an ancient technology but they became widespread following the Industrial Revolution. Industrial cranes were used to handle heavy machinery at the Nasmyth, Gaskell and Company (Bridgewater foundry) in the late 1830s. Hydraulic powered cranes became widely used in the late 19th century, especially at British ports. Some cities, such as London, had public utility hydraulic service networks to power. Steam cranes were also used in the late 19th century. Electric cranes, especially the overhead type, were introduce in factories at the end of the 19th century. Steam cranes were usually restricted to rails. Continuous track (caterpillar tread) was developed in the late 19th century.

The important categories of cranes are:

Overhead crane or bridge cranes-travel on a rail and have trolleys that move the hoist to any position inside the crane frame. Widely used in factories.

Mobile crane Usually gasoline or diesel powered and travel on wheels for on or off-road, rail or continuous track. They are widely used in construction, mining, excavation handling bulk materials.

Fixed crane In a fixed position but can usually rotate full circle. The most familiar example is the tower crane used to erect tall buildings.

PALLETIZATION

A U.S. airman operating a forklift. Pallets placed in rear of truck are moved around inside with a pallet jack (below). Where available pallets are loaded at loading docks which allow forklifts to drive on.

The handle on this pumpjack is the lever for a hydraulic jack, which can easily lift loads up to 2-1/2 tonnes, depending on rating. Commonly used in warehouses and in retail stores.

Handling goods on pallets was a significant improvement over using hand trucks or carrying sacks or boxes by hand and greatly speeded up loading and unloading of trucks, rail cars and ships. Pallets can be handled with pallet jacks or forklift trucks which began being used in industry in the 1930s and became widespread by the 1950s. Loading docks built to architectural standards allow trucks or rail cars to load and unload at the same elevation as the warehouse floor.

Piggyback rail

Piggyback is the transporting of trailers or entire trucks on rail cars, which is a more fuel efficient means of shipping and saves loading, unloading and sorting labor. Wagons had been carried on rail cars in the 19th century, with horses in separate cars. Trailers began being carried on rail cars in the U.S. in 1956. Piggyback was 1% of freight in 1958, rising to 15% in 1986.

Containerization

Either loading or unloading break bulk cargo on and off ships typically took several days. It was strenuous and somewhat dangerous work. Losses from damage and theft were high. The work was erratic and most longshoreman had a lot of unpaid idle time.

Sorting and keeping track of break bulk cargo was also time-consuming, and holding it in warehouses tied up capital.

Old style ports with warehouses were congested and many lacked efficient transportation infrastructure, adding to costs and delays in port.

By handling freight in standardized containers in compartmentalized ships, either loading or unloading could typically be accomplished in one day. Containers can be more efficiently filled than break bulk because containers can be stacked several high, doubling the freight capacity for a given size ship.

Loading and unloading labor for containers is a fraction of break bulk, and damage and theft are much lower. Also, many items shipped in containers require less packaging.

Containerization with small boxes was used in both world wars, particularly WW II, but became commercial in the late 1950s. Containerization left large numbers of warehouses at wharves in port cities vacant, freeing up land for other development.

Farm Machinery

What is agricultural machinery? Agricultural or Farm equipment is any type of machinery used on a farm to help with farming. Agricultural machinery is machinery used in agricultural farming or other agriculture. The best-known example of agricultural machinery is a tractor. Now let's get into details of farm machinery types and their importance.

Tractor

The tractor is commonly used farm machinery. Both commercial and subsistence farmers create use of a tractor for doing work. It is usually a wheeled vehicle with large wheels. Tractors generally do not do any farm work themselves rather the other farm equipment that does some farm works are attached to it. The main task is to pull those machines so that they will execute the tasks they are meant for. Those machines that cannot propel themselves are attached to a tractor. It can be used to haul trailers to be used for agricultural purposes. Tractors can be used to power machines that cannot power themselves. It is obtainable in various sizes depending on the type of machine to be attached to it.

A tractor is an important farm vehicle. Agricultural implements can be towed behind or mounted on the tractor and a tractor may provide a source of power if the implement is mechanized. A farm tractor is used for pushing agricultural machinery, for plowing, tilling, disking, harrowing, planting, and similar tasks. There are two types;

- Two-wheel tractor
- Tracked tractor or Caterpillar tractor

Sprayer

As implicit from the name, this kind of farm equipment is used to spray insecticides, pesticides, herbicides, fertilizers and any other products meant to be sprayed on the farm or plants. It is obtainable in various types. Some of the types contain agricultural aircraft-sprayer and blowers-prayer. Liquid or powdered insecticides or pesticides can be sprayed using this kind of machines. It is possible to use some sprayer to water the farm.

FIELD CULTIVATOR

As the name already tells you, this machine is particularly meant for cultivating the farm or making the soil softer for planting. It can be used to get rid of weeds or to aerate or make the soil softer shortly after the crops have started to germinate. Cultivators are helpful in farming today. Hardened soils are prepared softer for easy penetration of nutrients and water. They help in the preparation of the soil for planting and in eliminating the weeds from the plant.

Shredders and cutters

These type of machines meant for shredding and cutting unwanted vegetative growth and weeds.

Cotton harvester

Cotton harvester is a harvesting type of farming machinery but it is meant for the harvesting of cotton. They are usually available in two types, namely, pickers and strippers.

Seeders and planters

After the land has been cultivated, seeders and planters are used for sowing of the seeds or planting of the crops. With this machine, a grouping of seeds or crops can be planted within a few hours.

WHEEL TRACTOR-SCRAPPER

It is designed for earthmoving and for weeding purposes in a crop row of 1.5 widths and above. Weeds within this depth are removed with this apparatus. Removed weeds are cut into pieces by the pointed blade of the scraper and they are buried deep in the soil. When buried weeds decay, they attach to the green manure available to the soil.

Plough

It is particularly meant for plowing the soil before the seeds are sowed or crops are planted. The make the earth softer so that the root of the crops planted can penetrate the soil in explore of nutrients.

Baler

Baler is one of the most generally used farm machinery. Baler is used to cut hay and straw. It is capable of raking and binding bales with twine for easy transportation and storage.

Machinery used for soil cultivation

Cultivator: Cultivator equipment is a farm implement for stirring and pulverizing the soil before planting or to eliminate weeds and to aerate and loosen the soil after the crop has begun to grow. It is powered by a tractor and stirs the soil, generally to a greater depth than does the harrow (which is usually not powered). Many are equipped with hydraulic wings that fold up to make road travel simple and safer. Similar but very smaller machines are used for gardening.

Cultipacker: A cultipacker is a piece of agricultural tool that crushes dirt clods, removes air pockets, and presses down small stones, forming a smooth, firm seedbed. The term cultipacker is exclusively applied to ridged rollers, while the terms field roller or land roller may refer to either a smooth or a ridged roller. Some farmers treat the terms as mutually exclusive, but others treat the ridged tools as a class of field rollers.

Plough: A plow or plow is used for initial cultivation of soil in preparation for sowing seed. The main function of plowing is to turn over the upper layer of the soil, bringing fresh nutrients to the surface. It aerates the soil and allows it to hold moisture better. In modern use, a plowed field is normally left to dry out and is then harrowed before planting.

A Disc Plow – Farm Machinery Types.

Rotary tiller: It is also known as a rotavator, rotary hoe, power tiller, or rotary plow. It is a motorized cultivator that works the soil by means of rotating blades. Rotary tillers are either self-propelled or drawn as an addition behind a tractor.

Strip-till: Strip-till is a conservation system that uses for minimum tillage. It combines the soil drying and warming advantages of conventional tillage with the soil-protecting advantages of no-till by disturbing only the portion of the soil that is to contain the seed row. This tillage is performed with special tools and create multiple trips, depending on the strip-till implement used, and field conditions. Each row that has been strip-tilled is generally about eight to ten inches wide.

Harrows: Harrows equipment are implemented for cultivating the surface of the soil. It is different in its effect from the plow, which is used for deeper cultivation. Harrowing is often carried out on fields to follow the rough end left by plowing operations. The main purpose of this harrowing is to break up lumps of soil and to provide a finer finish, a good soil structure that is suitable for seeding and planting operations. Harrowing may be used to remove weeds and to cover seed after sowing. There are mainly three types of harrows: spike harrows, drag harrows, and disk harrows.

Machinery used for planting:

Seed drill: This is a device for placing the seeds in a continuous flow in furrows at a uniform rate and at controlled depth with or without the arrangement of covering them with soil.

This ensures that seeds will be distributed evenly. The seed drill sows the seeds at the accurate seeding rate and depth, ensuring that the seeds are covered by soil. This saves them from being eaten by animals or being dried up due to exposure to the sun.

Broadcast seeder: A broadcast seeder or spreader or fertilizer spreader is a tractor implement generally used for spreading seed, lime or fertilizer.

Seed

Machine.

Seed cum fertilizer drill: Seed drills are fitted with fertilizer dropping attachment; distribute the fertilizer uniformly on the ground. It is called a seed cum fertilizer drill. Such a drill has a large seed box which is divided lengthwise into 2 compartments, one for seed and another for fertilizers. Seed drill can be classified as (i) Bullock drawn (ii) Tractor drawn.

MACHINERY USED FOR FERTILIZING & PEST CONTROL

Manure spreader: A manure spreader is also known as muck spreader or a honey wagon is an agricultural machine used to distribute manure over a field as a fertilizer. Manure spreaders began as ground-driven units which can be pulled by a horse or team of horses.

Slurry tank: A slurry tank is a trailer with a tank and pump which can be used for slurry (a combination of manure and water) to fertilize the fields. If we fertilize fields, we get more crops at the end of the season.

Sprayer: A sprayer is a piece of tools that applies herbicides, pesticides, and fertilizers to agricultural crops. Sprayers range in size from man-portable units (typical backpacks with spray guns) to self-propelled units related to tractors.

Farm machinery types used for irrigation

Irrigation equipment: Varieties of irrigation implements are used for irrigation depending on the financing capability of the farm operator and suitability of the implement according to sources of water. In general centrifugal pump, submersible pump, sprinkler, drip irrigation, etc are used for irrigation function.

Fire sprinkler system: A fire sprinkler system is an active fire protection method, consisting of water provide a system, providing adequate pressure and flow rate to a water distribution piping structure, onto which fire sprinklers are connected.

Centre irrigation: Irrigation is the artificial application of water to the soil generally for assisting in growing crops. In crop production, it is mostly used in dry areas and in periods of rainfall shortfalls, but to protect plants against frost. Irrigation is the opposite procedure of drainage, which is the natural or artificial removal of surface and sub-surface water from a given area.

Harvesting and threshing equipment

Harvesting equipment is a mechanical device used for harvesting. There are several types of harvesting machines which are normally classified by crop. Below are some of the equipment used for harvesting;

Combine harvester: A combine harvester, or simply a combine, is a device that combines the tasks of harvesting, threshing, and cleaning grain crops. The main objective is the harvest of the crop; corn (maize), soybeans, flax (linseed), oats, wheat, rye and, etc.).

Sickle: A sickle tool is a curved shape, hand-held agricultural tool normally used for harvesting grain crop or cutting grass for hay.

Mower: Mower is a machine used to cut herbage crop or grasses. The mower can be power operated, manual operated or animal drawn.

Reaper: Reaper is a machine used to harvest cereals crops at ground level. These reapers are either power tiller operated or tractor operated. The field capacity of tractor operated reaper ranges 0.2 to 0.4 ha per hour.

Conveyor belt: A conveyor belt is the carrying medium of a belt conveyor structure (often shortened to belt conveyor). A belt conveyor system is one of the main types of conveyor systems. This system consists of two or more pulleys (sometimes referred to as drums), with an endless loop of moving medium and the conveyor belt that rotates about them. One or both of the pulleys are powered, moving the belt & the material on the belt forward. The powered pulley is known as the drive pulley while the unpowered pulley is called the idler pulley.

Crop harvesting machine: The mechanical device which harvests forage crops cultivated in upland or paddy field and forms roll bale simultaneously was developed, is termed as crop harvesting machinery. This machine comprises of traveling, reaping and a baler part.

CROP HARVESTER

Grain harvesting machine: This grain harvesting machine is used to harvest grains, the edible brans or fruit seeds of a cereal crop.

Root crop harvesting machine: Traditionally root crops are harvested by using diggers and digger-pickers. Nowadays, several machines are obtainable in the market. Modern sugar-beet harvester is one of the main popular examples of the root crop harvesting machine.

Threshers: Threshers or threshing machine is used for the separation of grain from stalks and husks.

Vegetable harvesting machine: Nowadays, many machines are available for the harvesting of vegetables. But, these 'vegetable harvesting machines', are quite common among the global vegetable farmers.

3

INTERNAL COMBUSTION ENGINES

The heat engine is a device in which heat energy of fuel after burning it properly is used to convert it into mechanical energy (mechanical work). The heat engine is an equipment which generates thermal energy and transforms it into mechanical energy. The heat energy is produced by the combustion of fuel. The heat engines are of two types :

1. EXTERNAL COMBUSTION ENGINE

In this type of engine the fuel is burnt outside the engine cylinder, in a device called boiler to produce hot gas or steam, which is used in the engine cylinder to develop the mechanical work. The line diagram of working principle of External combustion engine is shown in Fig 2.1. The best example for external combustion engine is steam engine. The thermal efficiency of external combustion engine is usually in the range of 15 to 20%, which is very low compared to the internal combustion engine.

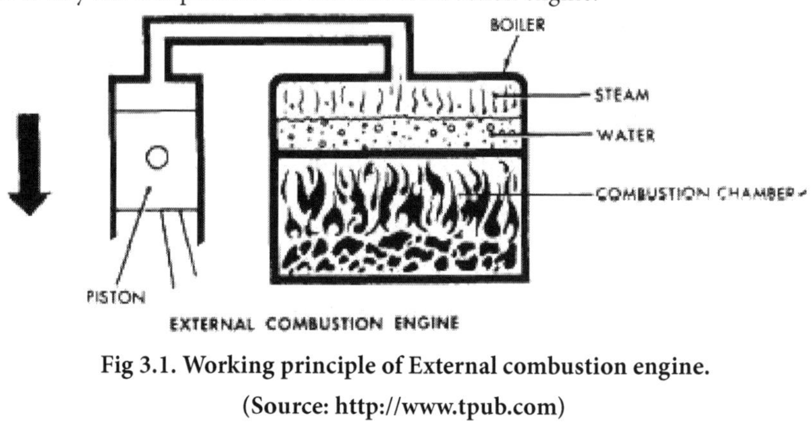

Fig 3.1. Working principle of External combustion engine.

(Source: http://www.tpub.com)

2. Internal Combustion Engine : In this type of engine the burning or combustion of the fuel takes place inside the engine cylinder. There are two ways in which combustion takes place in the cylinder:

i) By rapid explosion of air-fuel mixture within the cylinder, when it is ignited by a spark, is called constant volume combustion (C.V.C.).

ii) Combustion takes place by slow burning when the fuel is injected into highly compressed heated air contained in the cylinder. This is called constant pressure combustion (C.P.C.), because when the combustion takes place, the pressure in the cylinder is almost constant.

The power is developed during the power stroke. Most of the engines used in tractors, automobiles, combine harvesters, oil engines and traction vehicles are of this type. The line diagram of working principle of Internal combustion engine is shown in Fig. 3.2. The thermal efficiency of internal combustion engine is about 40%.

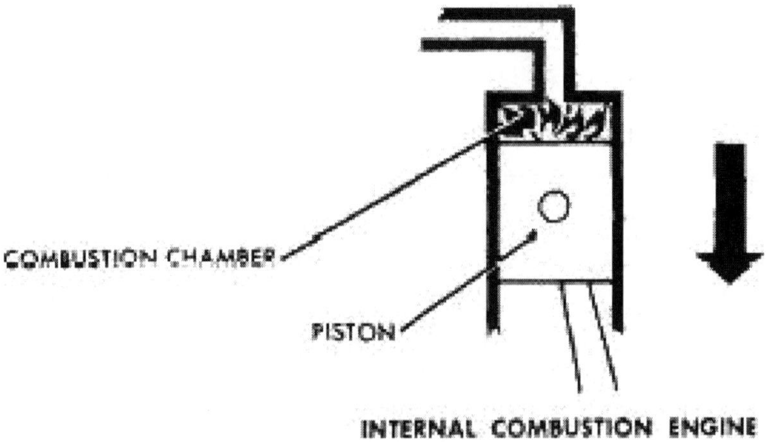

Fig. 3.2Working principle of Internal combustion engine

(Grace: http://www.tpub.com)

CLASSIFICATION OF INTERNAL COMBUSTION ENGINE (IC ENGINES)

The IC engines can be classified based on number of factors as follows

I. According to Number of Strokes :

1. Four Stroke Cycle Engine : In this type of engine, one complete combustion cycle is completed in four strokes of the piston or two revolution of the crankshaft. The principle of operation of Four stroke engine is shown in Fig. 3.3. The four events that take place inside the engine cylinder are:

1. Suction of air or air – fuel mixture in to the cylinder
2. Compression of the gases
3. Ignition of the fuel mixture to produce power

4. Exhaust of burnt gases

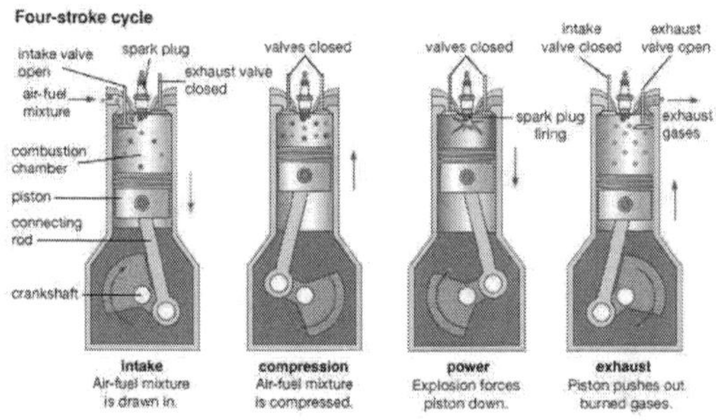

Fig. 3.3 Principle of operation of Four stroke engine

(Grace: http://www.britannica.com)

2. Two Stroke Cycle Engine: The two stroke cycle engine completes all the four events i.e. suction, compression, power or expansion and exhaust in two strokes (up stroke & down stroke) of piston or one revolution of crankshaft. During each stroke of the piston, two events takes place. The Principle of operation of Two stroke engine is shown in Fig. 3.4.

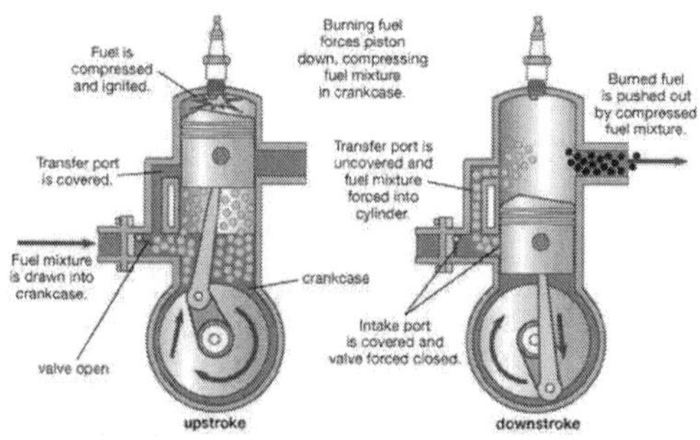

Fig. 3.4 Principle of operation of Two stroke engine

(Grace: http://tariqhafez.blogspot.com)

II. *According to thermodynamic working cycle:*

1. **Otto-cycle engine:** The heat is taken in at one constant volume and rejected at another constant volume of the cylinder. The combustion of the fuel takes place at constant volume. Petrol engine works on otto cycle.
2. **Diesel cycle engine:** In this cycle, the heat is taken at constant pressure and rejected at constant volume. The combustion of the fuel takes place at constant pressure. Diesel engine works on this cycle.
3. **Dual cycle engine:** In this type of engine, some part of the heat is taken at constant volume and the remaining part at constant pressure process. The heat is rejected at constant volume.

III. According to type of fuel used:

1. **Gas engine:** In this type of engine, the gaseous fuel like Compressed Natural Gas (CNG), coal gas, Bio gas, Liquid Petroleum Gas (LPG), producer gas is used to operate the engine. The engine is equipped with a special modified manifold system.
2. **Petrol engine:** In this type of engine, petrol is used as a fuel. A mixture of petrol and air enters the cylinder in a proper ratio during the suction stroke. The fuel mixture in vapour form is ignited with the help of a electric spark produced in the spark plug. Their method of ignition is called spark ignition ,Most of the two stroke and small engines use this system.
3. **Diesel engine:** In this type of engine, diesel is used as a fuel. Here, the air is compressed in the cylinder and atomized diesel is sprayed for ignition. The method of ignition is called compression ignition. All heavy vehicle use this system.

IV. According to number of cylinders:

1. Single cylinder engine: Here only one cylinder is present as shown in Plate 3.5. The power generated in the power stroke is used for functioning of other strokes of the engine.

Fig : 3.5 Single cylinder engine

(Grace: http://www.prakashagri.com)

1. Multi cylinder engine: Here more than one cylinder are present as shown in Plate 3.63 As per the firing order the power suction takes place in the respective cylinders.

Fig : 3.6 Multi cylinder engine

V. According to cylinder arrangement:

1. Horizontal cylinder engine: Most of the single cylinder engines are of this type of cylinder arrangement.
2. Vertical cylinder engine: Most of the present multiple cylinder engines have this type of cylinder arrangement.

VI. According to engine speed:

1. Low speed engine: The crankshaft speed of the engine is less than 100 rpm.
2. Medium speed engine: The crankshaft speed of the engine is in between 100 and 250 rpm.
3. High speed engine: Here the crankshaft speed of the engine is more than 250 rpm.

VII. According to cooling system used:

1. Air cooled engine: Here the cooling of the engine cylinder is done by air. In this case fins are provided on the cylinder as shown in Plate 3.7.

Fig : 3.7 Air cooled engine

2. Water cooled engine: Here cooling of the engine cylinder done by water circulated by a radiator and water pump as shown in plate 3.8.

Fig: 3.8 Water cooled engine

(Grace: http://www.indiamart.com)

ENGINE COMPONENTS

The internal combustion engine consists of the following engine components:

(i) Cylinder : It is a part of the engine which confines the expanding gases and forms the combustion space. It is the basic part of the engine. It provides space in which piston operates to suck the air or air-fuel mixture. The piston compresses the charge and the gas is allowed to expand in the cylinder, transmitting power for useful work. Cylinders are usually made of high grade cast iron.

(ii) Cylinder block : It is the solid casting which includes the cylinders and water jackets in case of water cooled engines. (cooling fins in the air cooled engines). Slated on engine and its components.

(iii)Cylinder head : It is a detachable portion of an engine which covers the cylinder and includes the combustion chamber, spark plug/atomizer and valves.

(iv) Cylinder liner or sleeve: It is a cylindrical lining either wet or dry which is inserted in the cylinder block in which the piston slides. Cylinder liner are fitted in the cylinder bore and they are easily replaceable. The overhauling and repairing of the engines, fitted with liners is easy and economical. The cylinder liners are classified as : (a) Dry liner and (b) Wet liner. The dry liner makes metal to metal contact with the cylinder block casing and do not come in contact with the cooling water. The wet liners come in contact with the cooling water.

(v) Piston : It is a cylindrical part closed at one end and open at the other end, which maintain a close sliding fit in the engine cylinder. The piston is shown in Plate 3.9. The piston is connected to the connecting rod by a piston pin. The force of the expanding gases against the closed end of the piston, forces the piston down in the cylinder. This causes the connecting rod to rotate the crankshaft connected rod is made of cast-iron. Cast iron is chosen due to its high compressive strength, low coefficient of expansion, resistance to high temperature, ease of casting and low cost. Aluminum and its alloys preferred mainly due to it lightness and easy dissipation of heat.

Fig: 3.9 Piston

(vi) Head (Crown) of piston : It is the top portion of the piston where the grooves are provided for inserting the piston rings and oil rings..

(vii) Skirt : It is that portion of the piston below the piston pin which is designed to absorb the side movements of the piston.

(viii) Piston ring : It is a split expansion ring, placed in the groove of the piston. Piston rings are fitted in the grooves, made in the piston. They are usually made of cast iron or pressed steel alloy. The function of the ring are as follows :

a) It forms a gas tight combustion chamber for all positions of piston.

b) It reduces contact area between cylinder wall and piston wall preventing friction losses and excessive wear.

c) It controls the cylinder lubrication.

d) It transmits the heat away from the piston to the cylinder walls. Piston rings are of two types : (a) Compression ring and (b) Oil ring

a. Compression ring : Compression rings are usually plain, single piece and are always placed in the grooves, nearest to the piston head.

b. **Oil ring:** Oil rings are grooved or slotted and are located either in the lowest groove above the piston pin or in a groove above the piston skirt. They control the distribution of lubrication oil in the cylinder and the piston. They prevent excessive oil consumption also. Oil ring is provided with small holes through which excess oil returns back to the crankcase chamber. Ring clearance is the gap at the joint of the ring, measured when the ring is inside the cylinder. The gap is usually 1 mm per 200 mm diameter of the piston. This clearance is necessary for expansion of the ring in heated condition, without which the ring can break or buckle. The piston ring and oil ring are shown in Plate 3.10.

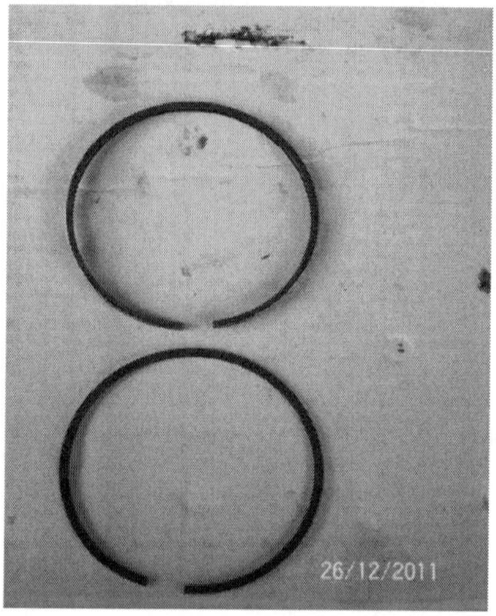

Fig : 3.10 Piston ring and oil ring

(ix) Piston Pin : It is also called as wrist pin or gudgeon pin. Piston pin is used to join the small end of the connecting rod to the piston. The piston pin is shown in Plate 3.11. It provides a flexible or hinge like connection between the piston and the connecting rod. It is usually made of case hardened alloy steel.

Fig. 3.11 Piston pin

(x) Connecting rod : It is special type of rod, small end of which is attached to the piston and the larger end to the crankshaft. It transmits the power of combustion to the crankshaft and makes it rotate continuously. It is usually made of drop forged steel. Its small end is fitted with bronze busing and big end is provided with split bearings. The connecting rod is shown in Plate 3.12.

Fig : 3.12 Connecting rod

(xi) Crankshaft : It is the main shaft of an engine which converts the reciprocating motion of the piston into rotary motion of the flywheel. The crankshaft of a four cylinder engine is shown in Plate 3.13. Usually the crankshaft is made of drop forged steel or cast steel. Crankshaft is provided with counter weights throughout its length to have counter balance of the unit. Split shell bearings are used as main bearings of the crankshaft. Crankshaft is subjected to bending as well as twisting from the connecting rod end.

Fig: 3.13 Crankshaft of a four cylinder engine

(xii) Flywheel : Flywheel is made of cast iron. The main functions of the flywheel are as follows:

(a) It stores energy during power stroke and returns back the same energy during the idle strokes, providing an uniform rotary motion by virtue of its inertia.

(b) The rear surface of the flywheel serves as one of the pressure surfaces for the clutch plate.

Engine timing marks are usually stamped on the flywheel, which helps in adjusting the timing of the engine. Two views of the flywheel are shown in plate 3.14.

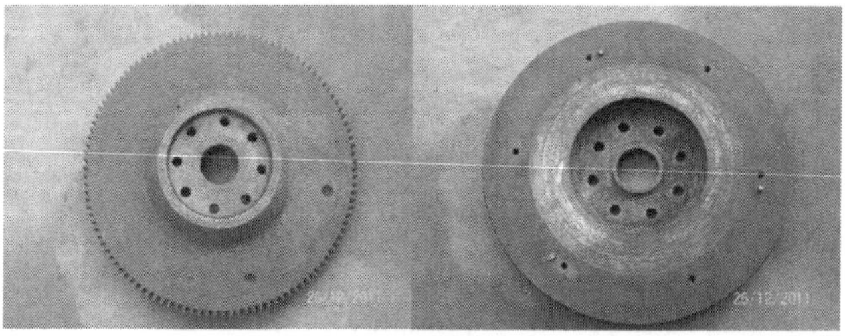

Fig: 3.14 Two views of the fly wheel

(xiii) Crankcase : The crankcase is that part of the engine which supports and encloses the crankshaft and camshaft. It provides a reservoir for the lubricating oil of the engine. It also serves as a mounting unit for accessories as like oil pump, oil filter, dynamo, self motor and other components. The upper portion of the crankcase is usually integral with cylinder block. The lower part of the crankcase is commonly called oil pan and is usually made of cast iron or cast aluminum. The crankcase of a four cylinder engine is shown in plate 3.15.

Fig : 3.15 Crankcase of four cylinder engine

(xiv) Camshaft : It is a shaft which raises and lowers the inlet and exhaust valves at proper time. Camshaft is driven by crankshaft by means of gears, chains or sprockets. The camshaft of a four cylinder, four stroke engine is shown in Plate 3.16. The speed of the camshaft is exactly half the speed of the crankshaft in four stroke engine. Camshaft operates the ignition timing mechanism, lubricating oil pump and fuel pump. It is mounted in the crankcase, parallel to the crankshaft.

Fig : 3.16 Cam shaft of a four cylinder engine

(xv) Timing gear : Timing gear is a combination of gears, one gear of which is mounted at one end of the camshaft and the other gear at the end of the crankshaft. Camshaft gear is bigger in size than that of the crankshaft gear and it has twice as many teeth as that of the crankshaft gear. For this reason, this gear is commonly called half time gear. Timing gear controls the timing of ignition, timing of opening and closing of valve as well as fuel injection timing. Plate 3.17 shows the timing gear and chain drive of an engine.

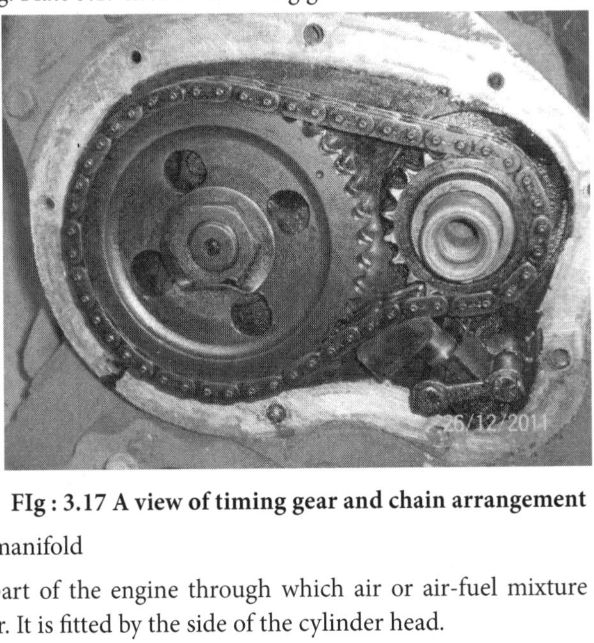

FIg : 3.17 A view of timing gear and chain arrangement

(xvi)Inlet manifold

It is that part of the engine through which air or air-fuel mixture enters into the engine cylinder. It is fitted by the side of the cylinder head.

(xvii) Exhaust manifold : It is that part of the engine through which exhaust gases go out of the engine cylinder. It is capable of withstanding high temperature of burnt gases. It is fitted by the side of the cylinder head.

The different parts of piston and engine parts along with their position in the engine are shown in Plates 3.18a and 3.18b.

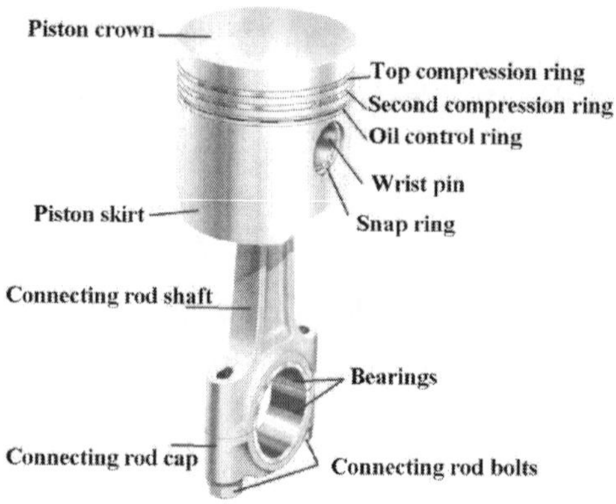

Fig: 3.18 a Different parts of the piston

(Grace: http://www.motorera.com)

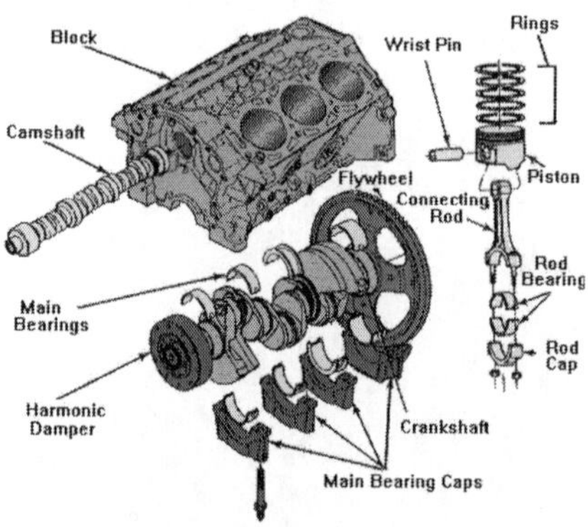

Fig : 3.18 b Different parts of the engine (Source: http://www.aa1car.com)

4

FOUR STROKE AND TWO STROKE ENGINE

PRINCIPLE OF I.C. ENGINE

A mixture of fuel with correct amount of air is exploded or burnt inside the engine cylinder which is closed at one end. As a result of the explosion, heat is released and this causes the pressure of the burning gases to increase, regulating the piston to move down in the cylinder. The movement of piston is transmitted to a crankshaft by a connecting rod, so that the crankshaft rotates the flywheel. To obtain continuous rotation of the crankshaft explosion has to be repeated. In addition to this, the burnt gases has to be expelled from the cylinder, fresh charges of fuel and air must be admitted and the piston must be moved back to its original position.

WORKING OF I.C. ENGINE

In IC engine the piston which is reciprocating in the cylinder is tight fit in the cylinder. Rings are inserted in the circumferential grooves of the piston to prevent leakage of gases from sides of the piston. Usually a cylinder is bored in a cylinder block and a gasket, made of copper sheet or asbestos is inserted between the cylinder and the cylinder head. The combustion space is provided at the top of the cylinder head where combustion takes place. There is a rod called connecting rod for connecting the piston and the crankshaft. A pin called gudgeon pin or wrist pin is provided for connecting the piston and the connecting rod of the engine. The end of the connecting rod which fits over the gudgeon pin is called small end of the connecting rod. The other end which fits over the crank pin is called big end of the connecting rod. The crankshaft rotates in main bearings which are fitted in the crankcase. A flywheel is provided at one end of the crankshaft for smoothening the uneven torque, produced by the engine. In I.C. engine the connecting rod converts the reciprocating motion of piston into rotary motion of the crankshaft. There is an oil sump at the bottom of the engine which contains lubricating oil for lubricating different parts of the engine. The cycle of internal combustion engine can be completed in two ways:

a. When the cycle is completed in two revolutions of the crankshaft or four strokes of the piston, it is called four stroke cycle engine.

b. When the cycle is completed in one revolution of the crankshaft or two strokes of the piston, it is called two stroke cycle engine.

1. Four Stroke Cycle Engine

In four stroke cycle engine, all the events taking place inside the engine cylinder are completed in four strokes of the piston. This engine has got valves for controlling the charges and exhaust gases. The opening and closing of the valve is controlled by cams, fitted on camshaft. The camshaft is driven by crankshaft with the help of suitable gears or chains. The camshaft runs at half the speed of the crankshaft. The events taking place in

I.C. engine are as follows :

I. Air or air-fuel mixture (charge) is taken in the cylinder.

II. The charge is compressed in the cylinder by the piston.

III. If charge is only air, the fuel is injected at the end of compression.

IV. The charge is ignited at a predetermined time under specified pressure inside the engine cylinder.

V. The power developed due to expansive forces of gases inside the cylinder is transferred to the crankshaft through the connecting rod.

VI. Exhaust gases go out of the cylinder at regular interval of time.

The complete cycle covers all these events in systematic manner. Four stroke cycle engines complete all these events in four strokes of the piston, whereas the two stroke cycle engine covers all these events in two strokes of the piston. The four strokes of the piston are:

I. Suction stroke II. Compression stroke

III. Power stroke or expansion stroke IV. Exhaust stroke

i. Suction /Induction stroke

During suction stroke (Fig. 4.1), only air or mixture of air and fuel are drawn inside the cylinder. The charge enters the engine through the inlet valve which remains open during admission of the charge. The exhaust valve remains closed during this stroke. The pressure in the engine cylinder is less than atmospheric pressure during this stroke due to which the charge is sucked in to the engine cylinder.

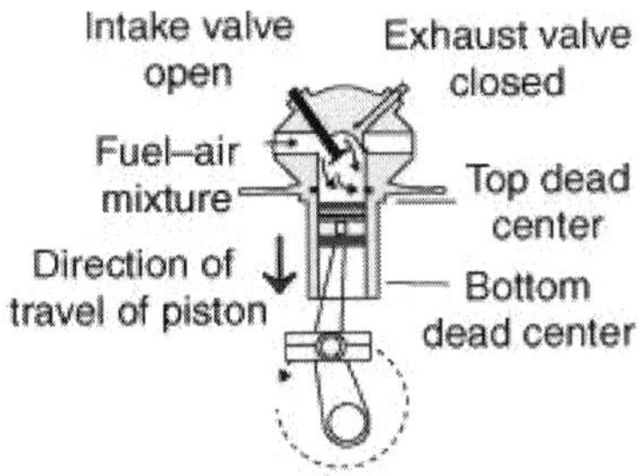

Induction stroke.

Fig. 4.1 Suction stroke

(Source: http://content.answcdn.com)

ii. Compression stroke

The charge taken in the cylinder is compressed by the piston during this stroke (Fig. 4.2). The entire charge of the cylinder is compressed to a small volume contained in the clearance volume of the cylinder. If only air is compressed in the cylinder (as in case of compression ignition engine), the fuel is injected at the end of the compression stroke. The ignition takes place due to high pressure and temperature. If the mixture of air and fuel is compressed in the cylinder (as in case of spark ignition engine) the mixture is ignited by spark plug. After ignition, tremendous amount of heat is generated, causing very high pressure in the cylinder which pushes the piston backward for useful work. Both valves are closed during this stroke.

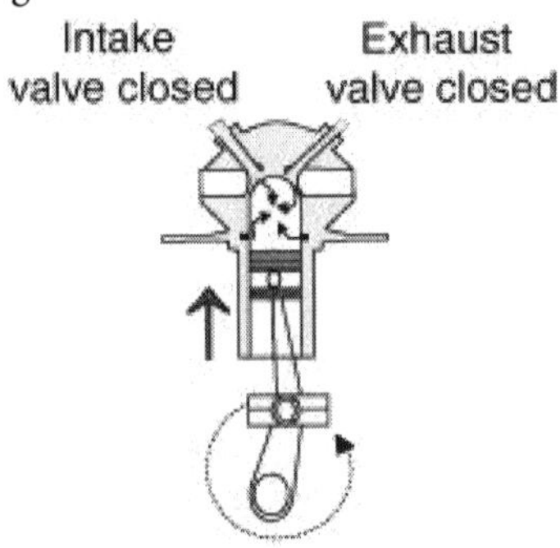

Compression stroke.

Fig. 4.2. Compression stroke

(Source: http://content.answcdn.com)

iii. Power Stroke

During power stroke (Fig. 4.3), the high pressure developed due to combustion of fuel causes the piston to be forced downward at regular intervals. The connecting rod with the help of crankshaft transmits the power to the transmission system for useful work. Both valves are closed during this stroke.

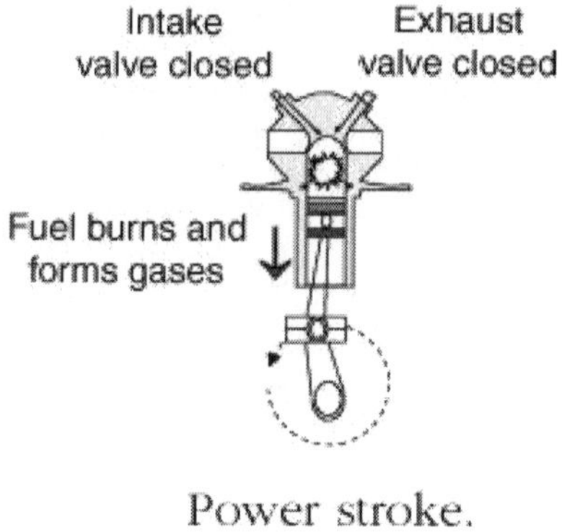

Fig. 4.3 Power stroke

(Source: http://content.answcdn.com)

iv. Exhaust stroke

Exhaust gases go out through exhaust valves during this stroke. All the burnt gases go out of the engine and the cylinder becomes ready to receive the fresh charge.

The inlet valve is closed and exhaust valve remains open during this stroke as shown in Fig. 4.4.

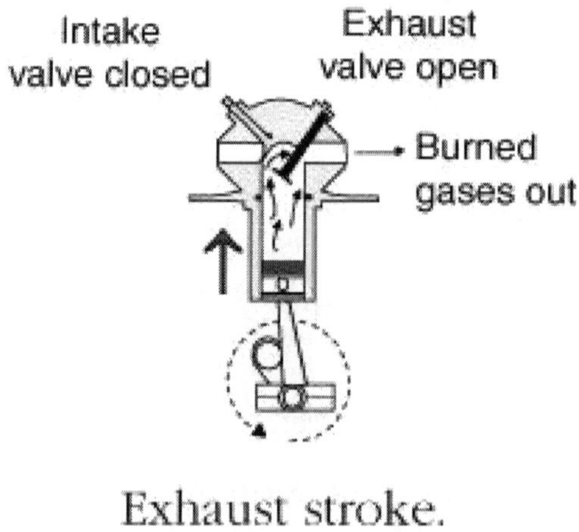

Fig. 4.4 Exhaust stroke

(Source: http://content.answcdn.com)

Thus it is found that out of four strokes, there is only one power stroke and three idle strokes. The power stroke supplies necessary momentum for remaining three stroke. The working of four stroke engine and the position of different parts are shown in Plate 4.5.

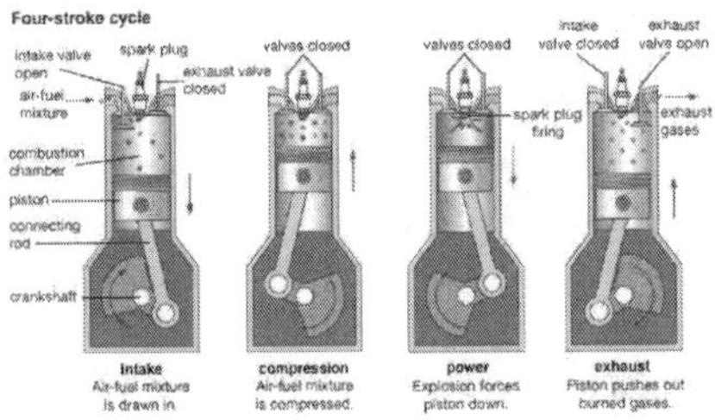

Fig: 4.5 Principle of operation of Four stroke engine

(Grace: http://www.britannica.com)

b. Two stroke cycle engine: In such engines, the whole sequence of events i.e., suctions, compression, power and exhaust are completed in two strokes of the piston or one complete revolution of the crankshaft. There is no valve in this type of engine. Gas movement takes place through ports in the cylinder. The crankcase of the engine is gas tight in which the crankshaft rotates.

i. First stroke (Suction and Compression)

When the piston moves up in the cylinder it covers two of the ports, the exhaust port and transfer port, which are normally almost opposite to each other. This traps a charge of fresh mixture in the cylinder and further upward movement of the piston compresses this charge. Further movement of the piston also uncovers a third port in the cylinder called transfer port and the charge is compressed. More fresh mixture is drawn through this port into the crankcase. Just before the end of this stroke, the mixture in the cylinder is ignited in the two stroke cycle. The first stroke is depicted in Plate 4.6.

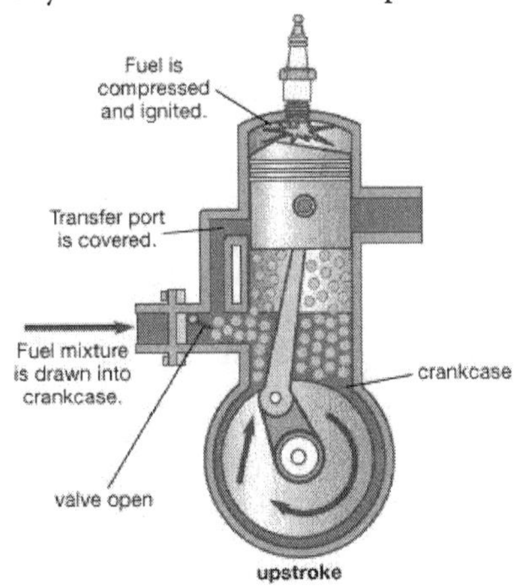

Fig. 4.6 First stroke in two stroke engine

(Source: http://3.bp.blogspot.com)

ii. Second stroke (Power and Exhaust)

The rise in pressure in the cylinder caused by the burning gases forces the piston to move down in the cylinder. When the piston goes down, it covers and closes the suction port, trapping the mixture drawn into the crankcase during the previous stroke then compressing it. Further downward movement of the piston uncovers first the exhaust port and then transfer port. This allows the burnt gases to flow out through exhaust port. Also the fresh mixture under pressure in the crankcase is transferred into the cylinder through transfer port during this stroke. Special shaped piston crown deflect the incoming mixture up around the cylinder so that it can help in driving out the exhaust gases. The Second stroke is depicted in Plate 4.7.

When the piston is at the top of its stroke, it is said to be at the top dead centre (TDC), when the piston is at the bottom of its stroke, it is said to be at its bottom dead centre (BDC). In two stroke cycle engine, both the sides of the piston are effective which is not in the case of four stroke cycle engine.

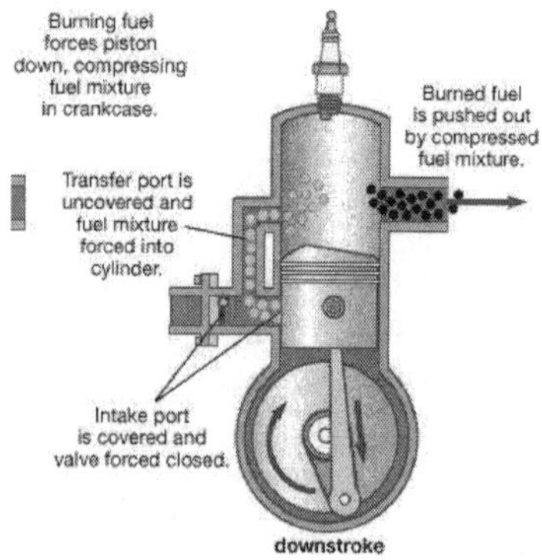

Fig : 4.7 Second stroke in two stroke engine

(Source: http://3.bp.blogspot.com)

Scavenging

The process of removal of burnt or exhaust gases from the engine cylinder is known as scavenging. Entire burnt gases do not go out in normal stroke, hence some type of blower or compressor is used to remove the exhaust gases in two stroke cycle engine. Comparison between two stroke and four stroke engine

Four stroke engine	**Two stroke engine**
One power stroke for every two revolutions of the crankshaft.	One power stroke for each revolution of the crankshaft.
There are inlet and exhaust valves in the engine.	There are inlet and exhaust ports instead of valves.

Crankcase is not fully closed and air tight.	Crankcase is fully closed and air tight.
Top of the piston compresses the charge.	Both sides of the piston compress the charge.
Size of the flywheel is comparatively larger.	Size of the flywheel is comparatively smaller.
Fuel is fully burnt.	Fuel is partially burnt.
Weight of engine per hp is high.	Weight of engine per hp is comparatively low.
Thermal efficiency is high.	Thermal efficiency is comparatively low.
Removal of exhaust gases easy.	Removal of exhaust gases comparatively difficult.
Torque produced is even.	Torque produced is less even.
All types of speed are possible (high and low).	Mostly high speed engines are there.
It can be operated in one direction only.	It can be operated in both directions (clockwise and counter clockwise).

5

SPARK AND COMPRESSION IGNITION ENGINES

The internal combustion engines are classified based on Spark ignition engine (petrol engine) and Compression ignition engine (diesel engine) method of ignition of the fuel in the engine cylinder.

SPARK IGNITION ENGINE (PETROL ENGINE)

In this type of engine, a device called carburetor, which atomizes and mixes fuel with air in correct proportion and send the mixture to the engine cylinder during the suction stroke. Some times these engines are also regarded as carburetor type engine.

It is an engine designed to convert chemical energy of heavier fuel into mechanical energy. The injected fuel is ignited by the heat of the air which is compressed by the piston within the cylinder. In this engine only, air is sucked into the engine cylinder during suction stroke.

Compression ignition engine (Diesel engine)

In compression ignition engines only air is drawn into the cylinder instead of mixture of air and fuel as in carburetor type engine. Compression of the air during the compression stroke makes the charge very hot, and increasing pressure and temperature. At the end of the compression stroke, diesel fuel is injected into the cylinder in atomised form through an injector, which ignites in the engine cylinder, causing rise in pressure which forces the piston to move downwards.

The basic elements of CI engine are same as that of spark ignition engine but the method of fuel introduction and ignition are different to a great extent. The engine has high compression ratio hence the air in the cylinder attains a very high temperature and pressure at the end of the compression stroke. At the end of the compression stroke, the fuel is sprayed into the cylinder through an atomizer (injectors). The cylinder contains air at high pressure and temperature; hence the fuel begins to burn as soon as the fuel reaches the cylinder in atomised form. Such engines are called compression ignition engines because the ignition of fuel takes place due to heat of compression. CI engine is equipped with fuel injection pump and injectors. The injectors protrude into the combustion space of the engine.

Special features of diesel engine

(1) Engine has high compression ratio ranging from 14:1 to 22:1.

(2) During compression stroke, the engine attains high pressure ranging from 30 to 45 kg/cm2 and high temperature of about 500°C.

(3) At the end of the compression stroke, fuel is injected into the cylinder through injectors (atomizers) at a very high pressure ranging from 120 to 200 kg/cm2.

(4) Ignition takes place due to heat of compression.

(5) Diesel engine has better slogging or lugging ability i.e. it maintains higher torque for a longer duration of time at a lower speed.

Comparison between Compression ignition engine and Petrol ignition engine

Compression ignition engine	**Petrol ignition engine**
It has got fuel injection pump and injector.	It has got carburetor, ignition coil and spark plug.
The compression ratio varies from 14:1 to 22:1	The compression ratio varies from 5:1 to 8:1.
It uses diesel oil as fuel.	It uses petrol (gasoline) or power kerosene as fuel.
Only air is sucked in cylinder in suction stroke.	Mixture of fuel and air is sucked in the cylinder in suction stroke.
Fuel is injected in combustion chamber where burning of fuel takes places due to heat of compression.	Air fuel mixture is compressed in the combustion chamber when it is ignited by an electric spark.
Thermal efficiency varies from 32 to 38%	Thermal efficiency varies from 25 to 32%
Engine weight per horse-power is high.	Engine
Operating cost is low.	Operating cost is high.
Compression pressure inside the cylinder varies from 35 to 45 kg/cm2 and temperature is about 500°C.	Compression pressure varies from 6 to 10 kg/cm2 and temperature is above 260°C.

6

COOLING SYSTEM

Fuel is burnt inside the cylinder of an internal combustion engine to produce power and hence large amount of heat is produced inside the engine cylinder. This raises the temperature of the cylinder, piston and valves. The temperature may reach 1600 oC, which is above the melting point of the engine parts in contact. The material strength is reduced even at 150 oC. Because of heat the lubricating oil is also gets oxidized. So the cooling system is necessary for the engine to remove the excess heat from the engine block and to maintain the temperature of the components within the limit. The cooling system keeps the engine running efficiently. It is estimated that about 40% of the total heat produced in the engine cylinder is passed to the atmosphere by the exhaust gases, 30% is removed by cooling system and only about 30% is used to produce useful power. Effect of high temperature in the engine

- Cylinder and piston may expand to such an extent that the piston would seize in the cylinder and stop the engine.
- Lubricating quality of the oil inside the cylinder would be reduced due to high temperature and resulting poor sucking of air in the cylinder
- Preignition of fuel mixture would take place and cause engine knocking as well as loss of power.
- For satisfactory performance of the engine, neither overheating nor over-cooling is desirable. Experiments have shown that best operating temperature of Internal combustion engine lies between 60°C to 93°C, depending upon types of engines and load conditions.

PURPOSE OF COOLING

- The main purpose of cooling are :
- To maintain optimum temperature of engine for efficient operation under all conditions.

- To dissipate surplus heat for protection of engine components like cylinder, cylinder head, piston, piston rings and valves.
- To maintain the lubricating property of the oil inside the engine cylinder for normal functioning of the engine.

METHODS OF COOLING THE ENGINE

There are two methods of cooling systems used in IC engines namely Air cooling system and Water cooling system.

1. *Air cooling system*

In this type of cooling system, air is used as a medium for removing heat from engine to atmosphere. Air cooled engines are those engines, in which heat is conducted from the working components of the engine to the atmosphere directly. In such engines, cylinders are generally not grouped in a block. Air cooling is usually employed for light vehicles.

Radial ribs or fins (Plate 6.1) are used to increase the external surface area of the cylinder. The length of these fins will be the greatest where the cylinder is the hottest, particularly near the cylinder head and will progressively reduce towards the crank case. Air-cooled engines are mounted on motorcycle frame. They are usually exposed to the surrounding atmosphere. They relay on natural air stream. Air circulates around the cylinders, head and crank case during forward movement of the vehicle.

Fig: 6.1 A view of fins provided on the cylinder

In some engines, the cylinder of an air cooled engine has fins to increase the area of contact with air for speedy cooling. The cylinder is normally enclosed in a sheet metal casing called Cowling. The flywheel has blades projecting from its face, so that it acts like a fan drawing air through a hole in the cowling and directing it around the cylinder fins. For maintenance of air cooling system, passage of air is kept clean. This is done by removing the cowling and cleaning out the dirt etc, by a stiff brush or compressed air. When separate fan is provided, the belt tension is to be checked and adjusted if necessary.

Advantages of air cooled engine

- It is simpler in design and construction.
- Water jackets, radiators, water pump, thermostat, pipes, hoses etc. are eliminated. It is more compact.
- Air-cooled engines operate well in both hot and cold climates.
- Air-cooled engines, rapidly reach their working temperature from cold. Air cooling is natural cooling and it is free from corrosive products.
- This system is more reliable and less maintenance is required.
- Air-cooled engines can operate at higher working temperatures than equivalent liquid cooled.
- Air cooled engines have no coolant leakage (or) problems It is comparatively lighter in weight.

DISADVANTAGES OF AIR COOLED ENGINE

- There is uneven cooling of the engine parts.
- Each cylinder has to be cast individually. Fin design and casting is difficult.
- Used only in small engines provided in brush cutter and weeder engines, portable engines, lawn mower engines, etc.,

2. *Water cooling system*

In this type of engine water or coolant (heavy water) is used as a cooling medium. These engines are called "water cooled engines". The liquid is circulated around the cylinders by means of jacket provided, to absorb heat from the cylinder walls and further the heated water is conducted through a radiator which helps in cooling the water as shown in Fig. 6.1.

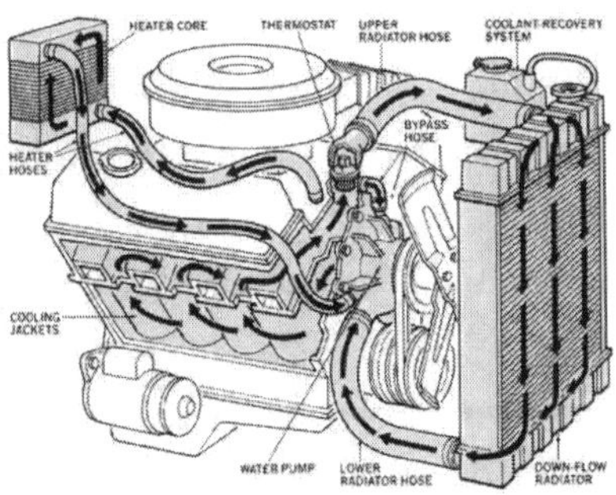

Fig. 6.2 Flow diagram of cooling liquid in the engine cooling system

(Courtesy: http://coolingtower-design.com)

There are four common methods of water cooling:

i. Open jacket or hopper method

ii. Thermo syphon method

iii. Forced circulation method

iv. Pressurized cooling

i. Open jacket or hopper method

In this method a hopper or a jacket is provided around the engine cylinder in which the cooling water or coolant is flowing. As long as the hopper contains water, the engine continues to operate satisfactorily. The hopper is large enough to run for several hours without refilling. A drain plug is provided for draining water. This system is not common in present days.

ii. Thermo syphon method

It consists of a radiator, water jacket, fan, temperature gauge and hose connections as shown in fig. 6.3. The system is based on the principle that heated water which surrounds the cylinder becomes lighter in weight and it rises up wards in liquid column. Hot water goes to the radiator, where it passes through tubes surrounded by air. Circulation of water takes place due to the density difference between the hot water in the engine jacket and cold water in the radiator. These two are connected at both sides i.e. at the top and the bottom by using the hose pipe. A fan is driven with the help of a V-belt connected to the crankshaft to suck air through cells of the radiator unit resulting in cooling. The disadvantage of the system is that circulation of water is greatly reduced by accumulation of scale or foreign matter in the passage and consequently it causes overheating of the engine.

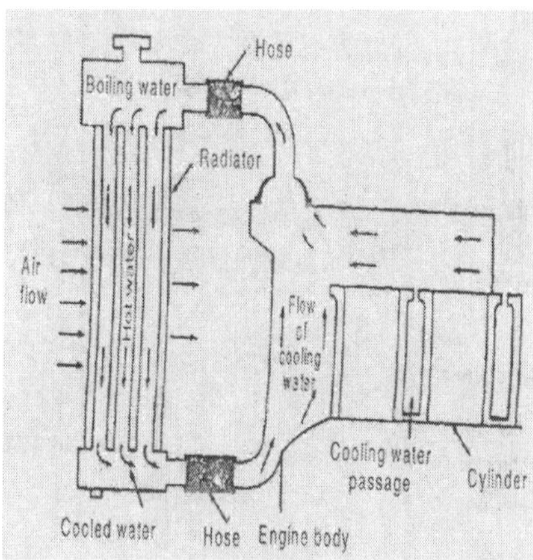

Fig. 6.3 Thermo syphon water cooling method

iii. Forced Circulation method

In this method, a water pump is used to force water from the radiator to the water jacket of the engine (Fig. 6.4). After circulating the entire run of water in jacket, water comes back to the radiator where it loses its heat by the process of conduction. To maintain the correct engine temperature, a thermostat valve is placed at the outer end of cylinder head. Cooling liquid is by-passed through the water jacket of the engine until the engine attains the desired temperature. Then thermostat valve opens and by-pass is closed, allowing the water to go to the radiator.

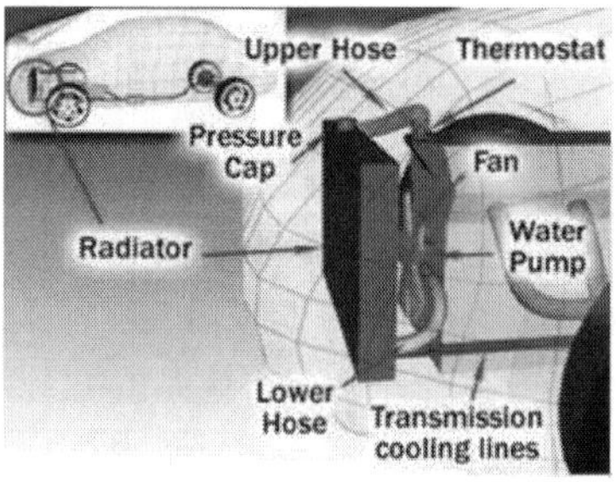

Fig. 6.4 Forced circulation system
(Courtesy: http://static.howstuffworks.com)

iv. Pressurized cooling:

In this type of cooling, the system is made tight enough to bear the additional pressure (Fig. 6.5). A special spring loaded cap for the radiator is provided to withstand the pressure. An overflow pipe is fitted above the spring loaded part of the cap. There is a vacuum relief disc, a spring and the pressure cap. When the engine cools, the water level in the radiator goes down and a vacuum is created. The vacuum operates the disc which is pushed down to save the radiator tubes from collapsing. With the increase in pressure, the boiling temperature of water increases from 100oC to 172oC. The higher water temperature gives more efficient engine performance and is useful at high altitudes.

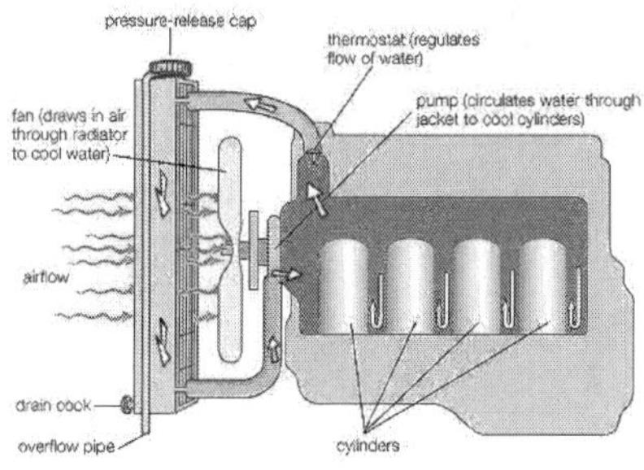

Fig. 6.5 Pressurized cooling system
(Courtesy: http://www.2carpros.com)

The different parts of the forced circulation method are explained below:

a. *Radiator*

The radiator transfers the heat absorbed by the water to the surrounding air. The radiator is flat and rectangular in shape as shown in plate 6.6. It consists of tubes which are made of copper or aluminum alloy. These tubes expose large surface area to the air stream. Fins are attached to these tubes. These fins improve the effectiveness of air convection heat dissipation.

Fig : 6.6 Radiator

b. *Cooling fan*

A fan (Plate 6.7) is positioned between the radiator and the engine and is driven by engine crankshaft. Fan is to increase the normal air movement across the radiator core tubes. The resultant air movement will provide a continuous air stream over the tubes and fins to dissipate the heat from the hot water which is circulated.

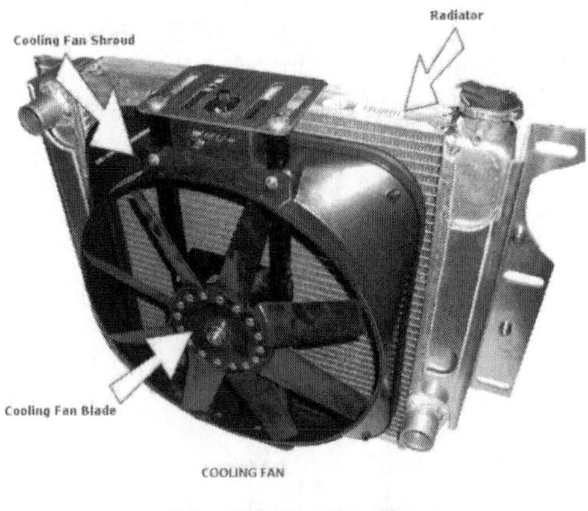

Fig: 6.7 Cooling fan
(Courtesy: http://www.2carpros.com)

C. Thermostat

The function of the thermostat is to regulate heat dissipation by controlling the rate of water flow through the radiator. Normally engines are designed to operate most efficiently over a temperature range between 80° C and 100°C. Thermostat (Plate 6.8) is placed in the outlet of cooling water from the engine. The thermostat has a valve, which opens only when the normal operating temperature of about 60°C is reaches by the water. The water flows from engine to upper part of the radiator through thermostat.

Fig : 6.8 Thermostat

(Courtesy: http://blog.autowarranties.com)

A centrifugal pump is used to speed up the rate of water circulation in the cooling system.

COOLING SYSTEM TROUBLES

In the engines, due to defective cooling system, several adverse effects are noticed, such as:

1. Overheating
2. Slow warm-up of the engine. Over heating is mostly due to:

a. Accumulation of rust and scale in the radiator and water jacket : Rust and scale accumulate in the radiator and water jacket. They restrict the circulation of water in the passage. Due to such restriction engine is overheated. Rust is caused by the oxidation of ferrous parts of the cooling system. It has got tendency to clog water passage and to insulate the iron part from the cooling liquid. Rust can be prevented by the use of suitable protective.

Scale is a hard deposit inside the water passage. It acts as a barrier to the flow of heat from the cylinder to the water.

b. Defective hose pipe : Sometimes, there is leakage in the hose pipe and consequently the water in the radiator is drained very quickly.

c. Defective thermostat : If the thermostat is stuck up in its seat due to some

distortion and resulting in Slow warm-up of the engine.

d. Defective water pump : Defective water pump disturbs the circulation of water in the system.

e. Loose fan belt : Loose fan belt causes slippage on the pulley and reduces effectiveness of cooling.

Care and maintenance of cooling system

- Clean and fresh water should be filled in the radiator.
- Salt free water should be used in the radiator as far as possible for prevention of scale formation.
- In some cold countries, the water in the radiator gets converted into ice and creates problems in initial starting of the engine. Hence ethyl alcohol is mixed with water to avoid ice formation.
- Rotten or soft hose pipe should not be used in the system.
- The tension of the fan belt should be checked very frequently. A loose belt will cause overheating, reduce charging current and high wear of the belt. If the belt

is very tight, it will cause wear on the pulley of pump and dynamo. The maximum permissible V-belt sag when applying average finger pressure is 15 mm. If it is more than this limit, the belt should be adjusted.

- Oil and grease should always be kept away from the belt. Greasy belts should be wiped clean.
- The bearing of the water pump should be lubricated regularly.

- Very hot engines should never be filled with cold water, to avoid fracture in the cylinder wall and the cylinder head.
- Radiator and water jackets should be flushed out with special air pressure guns.

Radiator can be cleaned by blowing air with compressor. This process will remove bugs, leaves and dirt from the radiator.

- The cooling system should be cleaned at periodically to remove rust and scale.
- The radiator cap should not be opened when the engine is hot.

The following suitable procedure may be followed to remove scale from the system.

One kg of washing soda and 0.5 kg of kerosene oil should be filled with fresh water and allowed to remain in the radiator for 8 to 10 hours. After this, the engine should be started and run on medium speed. When the engine has run for 15 to 20 minutes, the solution should be drained out and radiator should be flushed with clean water.

7

LUBRICATION SYSTEM

Internal combustion engines are made up of many moving parts. Due to continuous movement of two metallic surfaces over each other, there is continuous wearing of moving parts, generation of heat and loss of power in the engine. To prevent all these harmful effects the lubrication system is essential.

PURPOSE OF LUBRICATION SYSTEM

Reducing the friction: The primary function of the lubrication is to reduce the friction and wear and tear of the two rubbing parts. Two rubbing parts always produce friction and continuous friction produces heat, which causes wearing of the different parts of the engine and power loss. This friction can be reduced by proper lubrication. The lubricating oil forms a thin oil film between moving surface and avoids direct contact. This also reduces noise produced by the movement of two metal surfaces.

Cooling effect: The heat is generated by piston, cylinder and bearings is removed by lubricant to a great extent. Lubrication creates cooling effect on the engine parts and acts as coolant.

Sealing effect: The lubricant enters into the gap between the cylinder liner, piston and piston rings. Thus it acts as a seal preventing the leakage of the gas.

Cleaning effect: Lubrication keeps the engine parts clean by removing dirt or carbon from inside the engine.

TYPES OF LUBRICANTS

The lubricants are the different types of oils used for lubrication. Lubricants are obtained from animal fat, vegetables and minerals. The lubricants made of animal fat does not withstand much heat, it becomes gummy and waxy and hence not suitable for machines. Vegetable lubricants are obtained from seeds and plants such as cotton seed, olive oil, linseed oil and castor oil are used as lubricant in small machines. Mineral lubricants are most popular for engines and machines. It is obtained from crude petroleum, which is a natural resource. Mineral lubricants are less expensive and very much suited for internal combustion engine. Now days synthetic oil, which are produced from sand and coal such as poly alkaline and glycol are also used as lubricants.

There are different types of lubricants such as Solid lubricants (Graphite, Mica, Talc powder), Liquid lubricants (animal fat oil, vegetable oil, mineral oil) and Semi solid (Grease).

Properties of lubricating oil

The different properties of the lubricating oil :

Viscosity

It is the property of a lubricant by virtue of which it offers resistance to flow. The oil should be viscous enough to maintain a fluid film between the two mating surfaces. The viscosity is measured by an instrument called viscometer and expressed in terms of viscosity number. The unit of viscosity in SI system is Newton-second per square meter (N-s/m2) or Pascal seconds (Pas). Viscosity is inversely proportional to temperature i.e., as the temperature increases the viscosity decreases and vice versa.

FLASH POINT AND FIRE POINT

When the oil is heated, vapors are released. Flash point is the lowest temperature at which oil is to be heated until sufficient inflammable vapour come off which produces a momentary flash when brought to flame. When the vapours are released continuously and the flame persists for longer period, then that temperature is called fire point. A good lubricant has its flash and fire point above the temperature at which engine work, so that fire hazards are reduced.

Cloud point and Pour point

When lubricating oil is cooled, the temperature at which wax and other substance in the oil crystallize and separate out from oil is called Cloud point. The lowest temperature at which the oil ceases to flow when cooled is called Cloud point. Cloud point and Pour point indicates the suitability of lubricant for use in cold conditions. A good lubricant required to give service at low temperature should possess low pour point and cloud point.

Oiliness and Wettability

The ability of the lubricating oil to adhere to the surface is known as oiliness which depends upon on its wettability and surface tension. A good lubricant should have oiliness enough to adhere to the surface even at very high pressure.

Volatility: A good lubricant should have low volatility at working temperature.

Carbon residue: Lubricating oil has higher percentage of carbon in the combined form. A good lubricant should not deposit carbon when used at high temperature.

Engine lubrication system

The lubricating system of an engine is an arrangement of mechanism which maintains supply of lubricating oil to the rubbing surfaces of the engine at correct pressure and temperature. The engine parts which are to be lubricated are internal surfaces of cylinder walls, piston pin, piston rings, crankshaft, big and small end of the connecting rod, crankpin, valve operating mechanism, camshaft bearings, cooling fan, water pump and ignition mechanism.

There are three method of lubrication systems used in engines are:

- Petro-oil lubrication system.
- Splash lubrication system
- Forced feed or pressure lubrication system
- Combination Of splash and forced feed system

Petro-oil lubrication system

In this method of lubrication system, the lubricating oil is mixed with the petrol and fed in to the engine cylinder during the suction stroke. The droplets of the partials cause the lubricating effect in the engine cylinder. This method of lubrication is used in small engines like motorcycles and scooters. This method is not much effective in large engines.

The system of lubrication is used in scooters and motor cycles, particularly for two stroke engines about 3 to 6% of lubrication oil is added with petrol is the petrol tank. The petrol evaporates when the engine is working. The lubricating oil is left behind in the form of mist. The parts of the engine such as piston, cylinder walls and connecting rod are lubricated by being waited with the oil mist left behind.

When the added oil is less, there will be insufficient lubrication and even result in seizure of the engine. If the added oil is more, it will lead to excess exhaust smoke and carbon deposits in the cylinder exhaust port and spark plugs.

Splash lubrication system

In this system, the lubricating oil is filled in the sump or trough at the bottom of the crank case. Scoops (it is like a spoon) are attached to the big end of the connecting rod (Fig. 7.1). This pan receives its oil supply from the oil sump either by means of a gear pump or by gravity. A dipper is provided at the lower end of the connecting rod. This dipper dips into the oil trough and splashes oil out of the pan. The splashing action of oil maintains a fog or mist of oil that drenches the inner parts of the engine such as bearings, cylinder walls, pistons, piston pins, timing gears etc. The splash oil then drips back into the sump.

This system is commonly used in single cylinder engine with closed crankcase. For effective functioning of the engine, proper level of oil is maintained in the oil pan.

Lubrication depends largely upon the size of oil holes and clearances. This system is very effective if the oil is clean and undiluted. Its disadvantages are that lubrication is not very uniform and when the rings are worn, the oil passes the piston into combustion chamber, causing carbon deposition, blur smoke and spoiling the plugs. There is every possibility that oil may become very thin through crankcase dilution. The worn metal, dust and carbon may be collected in the oil chamber and be carried to different parts of the engine, causing wear and tear.

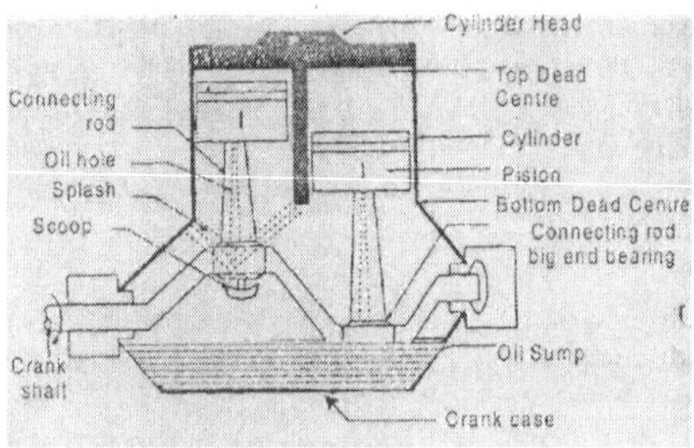

Fig. 7.1 Splash lubrication system

Forced feed or pressure lubrication system

In this system, the oil is pumped directly to the crankshaft, connecting rod, piston pin, timing gears and camshaft of the engine through suitable paths of oil as shown in fig. 7.2. Usually the oil first enters the main gallery, which may be a pipe or a channel in the crankcase casting. From this pipe, it goes to each of the main bearings through holes. From main bearings, it goes to big end bearings of connecting rod through drilled holes in the crankshaft. From there, it goes to lubricate the walls, pistons and rings. There is separate oil gallery to lubricate timing gears. Lubricating oil pump is a positive displacement pump, usually gear type or vane type. The oil also goes to valve stem and rocker arm shaft under pressure through an oil gallery. The excess oil comes back from the cylinder head to the crankcase. The pump discharges oil into oil pipes, oil galleries or ducts, leading to different parts of the engine. This system is commonly used on high speed multi cylinder engine in tractors, trucks and automobiles.

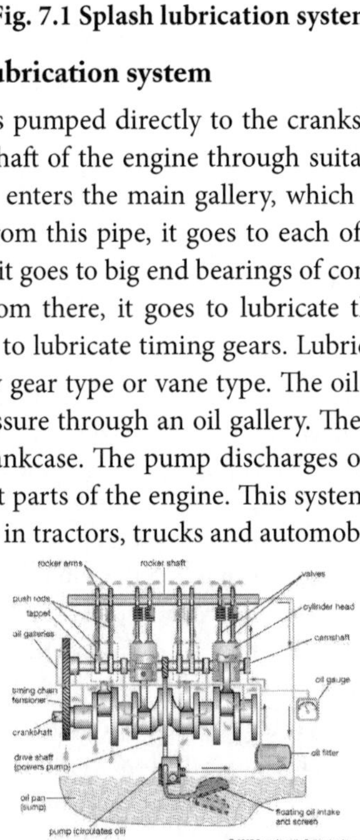

Fig. 7.2 Forced feed or pressure lubrication system Combination of splash and forced feed system

In this system, the engine component, which are subjected to very heavy load are lubricated under forced pressure, such as main bearing, connecting rod bearing and camshaft bearing. The rest of the parts like cylinder liners, cams, tappets etc. are lubricated by splashed oil.

PARTS OF LUBRICATION SYSTEM OIL PUM

Oil pump is usually a gear type pump, used to force oil into the oil pipe. The pump is driven by the camshaft of the engine. The lower end of the pump extends down into the crankcase which is covered with a screen to check foreign particles. A portion of the oil is forced to particles. A portion of the oil is forced to the oil filter and the remaining oil goes to lubricate various parts of the engine. An oil pressure gauge fitted in the line, indicates the oil pressure in the lubricating system. About 3 kg / cm2 pressure is developed in the lubrication system of a tractor engine. If the oil pressure gauge indicates low pressure in the line, there is some defect in the system which must be checked immediately. Lubricating oil pump is a positive displacement pump.

Troubles in lubrication system

There are some problems which are commonly faced in lubrication system such as: Excessive oil goes to combustion, low oil pressure and excessive oil pressure.

1. **Excessive oil consumption:** The reasons for excessive oil consumption are :

(a) More oil goes to combustion chamber and gets burnt

(b) Some leakage occurs in some part of the line and

(c) Loss of oil in form of vapour through ventilating system.

Oil can enter the combustion chamber through rings and cylinder walls, worn piston rings and worn bearings.

2. **Low oil pressure :** Low oil pressure can result due to :

(a) Weak relief valve spring

(b) Worn oil pump

(c) Cracked oil line

(d) Obstruction in the oil lines

(e) Very thin oil and

(f) Worn out bearings

Care should be taken to remove these defects as far as possible to increase the oil pressure in the lubricating system. Sometimes defective oil pressure indicator shows low oil pressure. This should be checked.

3. **Excessive oil pressure** : Excessive oil pressure may result due to:

(a) stuck relief valve

(b) strong valve spring

(c) clogged oil line and

(d) very heavy oil

These defects should be removed to reduce the excessive oil pressure in the lubricating system. Sometimes defective oil pressure indicator records high oil pressure. Care should be taken to check this defect.

CARE AND MAINTENANCE OF LUBRICATION SYSTEM

The following care and maintenance should be taken to work lubrication system properly

1. A good design of oil circulation system should be chosen.
2. Correct grade of lubricant ensures long and trouble free service.
3. Oil should be maintained at desired level in the oil chamber.
4. Oil should be cleaned regularly and after specified period of use, old filters should be replaced by new filters.
5. Connections, peeping, valves and pressure gauge should be checked regularly.
6. Oil should be changed regularly after specified interval of time. Before putting the new oil, the crankcase should be cleaned and flushed well with a flushing oil.
7. Precautions should be taken to keep the oil free from dust and water.

8

POWER TRANSMISSION SYSTEM

Transmission is a speed varying mechanism, equipped with several gears. It may be called a sequence of gears and shafts, through which the engine power is transmitted to the tractor wheels. The system consists of various devices, which cause forward and backward movement of tractor to suit different field condition. The complete path of power from the engine to the wheels is called Power train.

FUNCTIONS OF POWER TRANSMISSION SYSTEM

- The main functions of the power transmission system are:
- To transmit power from the engine to the rear wheels of the tractor. To make reduced speed available, to rear wheels of the tractor.
- To alter the ratio of wheel speed and engine speed in order to suit the field conditions.
- To transmit power through right angle drive, because the crankshaft and rear axle are normally at right angles to each other.
- The power transmission system consists of clutch, transmission gears, differential, final drive, rear axle and rear wheels as shown in Fig. 8.1. Combination of all these components are responsible for transmission of power from crankshaft to the rear wheels. These components are explained in the following sections.

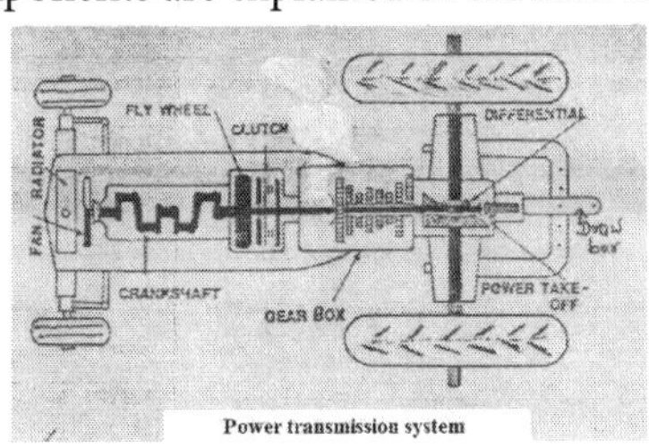

Fig. 8.1 Power transmission system

CLUTCH

Clutch is a device, used to connect and disconnect the tractor engine from the transmission gears and drive wheels. Clutch transmits power by means of friction between driving members and driven members.

Why clutch in a tractor is essential?

- Engine needs cranking by any suitable device. For easy cranking, the engine is disconnected from the rest of the transmission unit by a suitable clutch. After starting the engine, the clutch is engaged to transmit power from the engine to the gear box.
- In order to change the gears, the gear box must be kept free from the engine power, otherwise the gear teeth will be damaged and engagement of gear will not be perfect. This work is done by a clutch.
- When the belt pulley of the tractor works in the field it needs to be stopped without stopping the engine. This is done by a clutch.

The important essential features of a good clutch are:

- It should have good ability of taking load.
- It should have higher capacity to transmit maximum power without slipping. Friction surface should be highly resistant to heat effect.
- The control by hand lever or pedal lever should be easy.

Types of clutch: The clutches used in engines are of three types namely, Friction clutch, Dog clutch and Fluid coupling. Friction clutch is most popular in four wheel tractors. Dog clutch is mostly used in power tillers. Fluid clutch is used in some type of tractors in these days.

Gear and gear box

A tractor engine runs at high speed, but the rear wheel of the tractor requires power at low speed with high torque. Hence it is very much essential to reduce the engine speed and increase the torque available at the rear wheels of the tractor.

The power available is given by

$$\text{Power} = \frac{2\pi NT}{60 \times 1000} \quad \text{Watt}$$

where,

T - Torque in Newton-metre (N-m)

N - Speed in rpm

If the engine power is constant, it is obvious that for higher torque at wheels, low speed is required and vice versa. So the gear box (Plate 8.1) is fitted between engine and rear wheel for variable torque and speed. This is done by suitable design of gear and shafts. Speed varies according to the field requirements and hence a number of gear ratios are provided to suit the varying conditions. Gears are usually made of alloy steel.

Fig: 8.2 Gear box of a tractor

There are different types of gear box used in tractors such as Sliding mesh, Constant mesh and Synchromesh

Differential unit and final drive

Differential unit is a special arrangement of gears to permit one of the rear wheels of the tractor to rotate slower or faster than the other. While turning the tractor on a curved path, the inner wheel has to travel lesser distance than the outer wheel. The inner wheel requires lesser power than the outer wheel. This condition is fulfilled by differential unit, which permits one of the rear wheel of the tractor to move faster than the other at the turning point. The output shaft coming from the gear box is provided with a bevel pinion at the end of the shaft. The bevel pinion is in mesh with a large bevel wheel known as Crown wheel.

The main functions of Crown wheel assembly are to transmit power through right angle drive to suit the tractor wheels and to reduce the speed of rotation.

The Differential unit consists of different parts like differential casing, differential pinion, crown wheel, half shaft and bevel gear as shown in Fig. 8.2 The differential casing is rigidly attached with the crown wheel and moves like one unit. Two pinions are provided inside the differential casing, such that they are carried round by the crown wheel but they are free to rotate also on their own shaft or stud. There are two or more bevel gears in mesh with differential pinion. One bevel are two or more bevel gears in mesh with differential pinion. One bevel pinion is at the end of each half shaft, which goes to the tractor rear wheel. Thus instead of crown wheel being keyed directly to a solid shaft between the tractor wheels, the drive is taken back from the indirect route through differential casing, differential pinion and half shaft of the tractor. When the tractor is moving in a straight line, the differential pinion do not rotate on the stub shaft but are solid with the differential casing. They drive the two bevel gears at the same speed and in the same direction as the casing and the crown wheel.

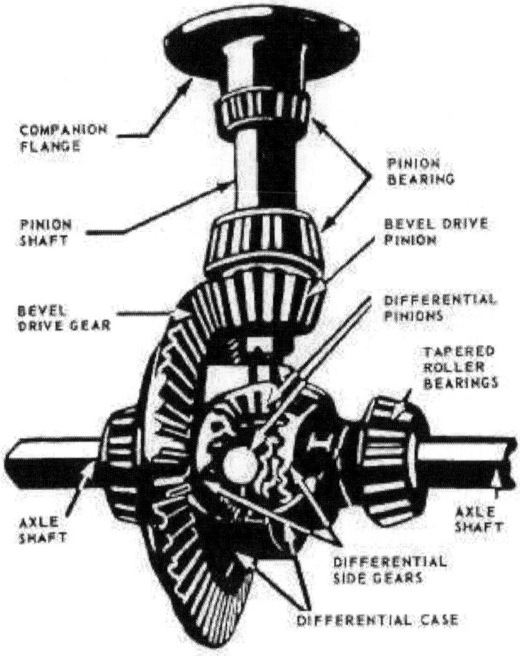

Fig. 8.3 Differential unit

Differential lock

Differential lock is a device to join both half axles of the tractor so that even if one wheel is under less resistance, the tractor comes out from the mud etc. as both wheels move with the same speed and apply equal traction.

FINAL DRIVE

Final drive is a gear reduction unit in the power trains between the Differential and the Drive wheels. Final drive transmits the power finally to the rear axle and the wheels. The tractor rear wheels are not directly attached to the half shafts but the drive is taken through a pair of spur gears. Each half shaft terminates in a small gear which meshes with a large gear called Bull gear. The Bull gear is mounted on the shaft carrying the tractor rear wheel. The device for final speed reduction, suitable for tractor rear wheel is known as Final drive mechanism.

9

ENGINE PERFORMANCE

The different terms used in the engine system are :

Bore: Bore is the diameter of the engine cylinder expressed in cm or metre. Stroke: It is the linear distance traveled by the piston from Top dead centre (TDC) to Bottom dead centre (BDC) and the distance travelled is called stroke length, expressed in cm or metre.

Stroke-bore ratio: The ratio of length of stroke (L) and diameter of bore (D) of the cylinder is called stroke-bore ratio (L/D). In general, this ratio varies from 1 to

1.45 and for tractor engines, this ratio is about 1.25.

Swept volume: It is the volume displaced by one stroke of the piston.

Swept volume = A x L .... m3 Where,

A - Cross sectional area of piston in m2 L - Stroke length in m.

Compression ratio: It is the ratio of the volume of the charge at the beginning of the compression stroke to the volume of the charge at the end of compression stroke, i.e. ratio of total cylinder volume to clearance volume.

The compression ratio of diesel engine varies from 14:1 to 22:1 and that of spark ignition engine (carburetor type engine) varies from 4:1 to 8:1.

Indicated power (ip)

It is the total power generated in the engine cylinder by the combustion of the fuel and received by the piston. It is the power developed in a cylinder without friction or auxiliary unit. In SI system, the indicated power is expressed in terms of Kilowatts (kW) and the formula for determining is given by

$$ip = \frac{PLAN}{60 \times 1012} \times \frac{n}{2} \quad \text{for four stroke engine}$$

PLAN x n

ip = ------------- ----   for two stroke engine

60 x 1012

Where,   P = mean effective pressure, Pa or N/m2 L = stroke length, mm

A = cross sectional area of piston, mm2 N = engine speed, rpm

n = number of cylinders

Brake power (bp): It is the power delivered by the engine and is available at the end of the crankshaft. It is measured by using a suitable dynamometer.

Mechanical efficiency (ηmech) : It is the ratio of brake power to indicated power expressed in percentage.

Belt power: It is the power of the engine, measured at the end of a suitable belt, receiving drive from the PTO shaft of the tractor.

Power take-off horse power (Pto power) : It is the power delivered by a tractor through its PTO shaft. In general, the belt and PTO horse power of a tractor will approximately be the same.

Drawbar power (db): It is the power of a tractor measured at the end of the drawbar. It is that power which is available for pulling loads at the draw bar.

Frictional power (fp) : It is the power required to run the engine at a given speed without producing any useful work. It represents the friction and pumping losses of an engine. It is the power lost due to the friction between the rubbing parts of the engine.

ip = bp + fp

Mean effective pressure (mep): It is the average pressure during the power stroke minus the average pressure during other strokes. This pressure actually forces the piston down during the power stroke.

Volumetric efficiency: It is the ratio of actual weight of air introduced by the engine on the suction stroke to the theoretical weight of air that should have been introduced by filling the piston displacement volume with air at atmospheric pressure and temperature.

Torque : A turning effect due to force applied on some point is called torque (T).

T = F x r   Kg-m or N-m

Where,

F = force in Kg or N

r = distance of the force from the centre of the shaft in meters.

Piston displacement: It is the volume displaced by one stroke (L) of the piston. It is also known as swept volume.

Piston displacement = A x L m3

Where,

A = cross sectional area in m2 L = length of stroke in meters.

Displacement volume: It is the total swept volume of all pistons during power strokes occurring in one minute.

Displacement volume = A x L x N x n/2 (for 4 stroke engine) Displacement volume = A x L x N X n/1 (for 2 stroke engine)

Where,

A = cross sectional area of piston in m2 L = stroke length in meters

N = speed in rpm

n = number of cylinders

Piston speed (Sp): It is the total length of travel of the piston in a cylinder in one minute.

Sp = 2 x Lx N

Specific fuel consumption: It is the quantity of fuel consumed per kW– hr. in an engine. The unit of specific fuel consumption is Kg/ kW-hr.

BRAKE DYNAMOMETER

The brake dynamometer is used to measure the brake power of the tractor engine. Different types of dynamometer are used to measure the power.

Problem 1 : Calculate the brake power and mechanical efficiency of a 4 stroke 4 cylinder internal combustion engine, having following dimensions: Cylinder bore, D = 125 mm, Stroke length, L = 150mm, Crankshaft speed, N = 1000 mm rev

/min, Friction power = 20 kW, Mean effective pressure, P = 700 kPa, No. of cylinder, x = 4.

Solution:

PLAN n

Indicated power (ip) = -------------- x ------- kW

60 x 1012 2

Where,

Pressure P = Pascal = N/m2 L = Length of stroke, mm

A = Cross sectional area, mm2 N = Speed, rpm.

n = Number of cylinder

700 x 1000 x 150 x ∏ x 125 x 125 x 1000 x 4

Indicated power (ip) =

60 x 1012 x 4 x 2

Indicated power (ip) = 42. 95 kW

Brake power = Indicated power – friction power

= 42.95 – 20

Brake power = 22. 95 kW

Mechanical efficiency, ηmech = BP/IP= 22.55 x 100 /42.55 =52.99%

Mechanical efficiency, ηmech = 52.99%

Problem 2 : Calculate the brake power of a 2 cylinder 4 stroke cycle I.C. engine 12 X 15 cm. The mean effective pressure is 700 k Pa and speed of crankshaft is 1200 revolutions per minute. The mechanical efficiency is 75%.

Solution:

Bore diameter, D = 120 mm Stroke length, L = 150 mm Mep, P = 700 k Pa

Speed, N = 1200 rpm. No. of cylinder, n = 2

We know that,

PLAN n

Indicated power (ip) = --------------- x ------- kW

60 x 1012 2

Where,

P - Pressure, Pascal

L - Length of stroke, mm

A - Cross sectional area, mm2 N - Speed, rpm.

n - Number of cylinder

700 x 1000 x 150 Π 120 x 120 x 1200 x 2

Indicated power (ip) = x x

60 x 1012 4 2

Indicated power (ip) = 23.75 kW

Brake power = Indicated x mechanical efficiency Brake power = 23.75 X 75/100

Brake power = 17.81 kW

Problem 3 : A four cylinder four stroke engine having cylinder bore 7.5 cm and stroke length 10 cm develops 15 kW at 1650 rev / min. Assuring a mechanical efficiency of 85%, find indicated power and mean effective pressure.

Solution :

Given,

Bore diameter, D = 7.5 cm = 75 mm Stroke length, L = 10 cm = 100 mm Brake power, bp = 15 kW

Speed. N = 1650 rpm . ηmech = 85 %

Number of cylinder, n = 4

Brake power 15

Indicated power (ip) = ------------------ = ------- =17.64 kW

ηmech 0.85

PLAN X

Indicated power (ip) = -------------- x ------- kW

60 x 1012 2

P x 100 Π 75 x 75 x 1650 x 4

17.64 = x x

60 x 1012 4 2

Mean effective pressure, P = 725. 97 kPa

Problem 4 : Determine the power and compression ratio of the engine, working on four stroke cycle principle. When it runs at 400 rev / min, the volume at the beginning of the compression is 0.032 m3 and at the end of compression is 0.005 m3. The work done per cycle is 650 kg-m.

Solution :

0.032 Compression ratio = = 6.4

0.005

Compression ratio = 6.4 : 1

400 10

Total cycles /sec. = ----------- = -----

2 X 60 3

Work done in 10/3 cycles / sec. = 10 x 650 kg- m / s3

= 2166.66 kg-m /s

= 2199.66 X 9.8 N-m /s

= 21233.26 watts.

Power of engine = 21.23 kW

10

TRACTORS AND POWER TILLERS

Tractor is a self propelled machine used either for pulling or pushing and stationary work. The word tractor is derived from two words traction and motor. Tractor is a machine which is used for traction purpose. Tractor can be used for different job by attaching implements, tools, as single unit. The tractor is farmer's movable power house. Tractor is a self propelled power unit having wheels or tracks for operating agricultural implements and machines including trailers. Tractor engine is used as a prime mover for active tools and stationary farm operations through power take-off shaft or belt pulley.

TRACTOR DEVELOPMENT

The present tractor is a result of gradual development of machines in different stages.

1890	Tractor engine invented by George Harris of Chicago, and used the world first tractor
1906	First gasoline tractor developed by Charles Hart from IOWA
1915-19	PTO – Power take off was introduced in tractor
1920-24	All purpose tractor developed
1936-37	Diesel engine and pneumatic tyres were used for tractors
1950-60	Large scale manufacturing of diesel tractors
1937-41	Hydraulic system and three point linkage developed
1960-61	Tractor manufacturing started in India by first manufacturer M/s. Eicher Good Earth.
1962-70	Manufacturers like Tractor and Farm equipment, Hindustan tractors, Escorts tractors, and International harvesters started work during this period

1971	Escort tractor limited started producing Ford Tractor
1973	Manufacturing of HMT tractors started
1974	Pitti & Kirlosker tractors were started
1975	Harsha tractors were started
1981	Auto tractors were developed
1982	Universal tractors were established
1983-08	GTCL tractors, Bajaj, Mahindra, Swaraja, Hindustan, Sonalika, Johndeer,
	PTL, VST and many other companies started manufacturing tractor in India.

CLASSIFICATION OF TRACTOR

The tractors are classified into three classes on the basis of type of construction, type of drive and purpose.

Based on type of construction, they are classified as

(a) **Riding type tractors** – Tractors in which a driver can sit and drive e.g., General purpose four wheel tractors.

(b) **Walking type tractors** – Tractors with which the operator walks along e.g., garden tractors, power tillers.

Based on type of drive :

(a) Track type tractors – In this type of tractors, instead of wheels; one track is fitted on either side. This track gets drive from the sprocket run by rear axle shaft. To steer the tractor, there is not steering gear fitted. The tractor is steered by applying brakes to one side of the track with the other track moving. These are used for bulldozing or land clearing work.

(b) Wheel type of tractors – These are most commonly used agricultural tractors. They can run fast and wheel tyres absorb a certain amount of field shocks also. These can be further divided as :

(1) Two-wheel tractors – These tractors are used for small farms, hilly area and gardening purposes and are called power tillers.

(2) Three-wheel tractors – These tractors were very popular 15 years back but now its place has been taken by four wheel tractors. These tractors had single or dual wheel fitted at the front end in the centre and were considered good for negotiable shorter turns.

(3) Four-wheel tractors – These are most commonly used tractors in the country. These are also known as all purpose tractors. On the basis of available power, these have been classified as

(1) Small tractors – 15 to 25 hp.

(2) Medium tractors – 25 to 45 hp.

(3) Large tractors – more than 45 hp.

Based on purpose :

(a) Utility tractors – It is a specific purpose field tractor and is designed for ploughing driving any other equipment through its P.T.O. drive. It is not being manufactured at present in the country.

(b) All purpose tractor – It is designed in such a way that it can meet practically all the demands for agricultural purposes such as ploughing, harrowing, leveling, pulling, seed drill, operating threshers, and pumps through its P.T.O. These are provided with three point linkages.

(c) Orchard type tractors - These are special types of tractors used in orchards. These are made very high in height so that driver while sitting on the seat, the operations on the trees could be performed. No part of the tractor is protruded outside so that tractor can go easily in between trees safely.

(d) Garden tractors - These tractors are in the range of 1 to 10 hp and are very small in construction. These are mostly used for kitchen or vegetable gardens.

(e) Rotary Tillers - These are walking type of tractors and are used in small fields or on hills where fields are very small and are at different levels where ordinary tractors cannot work efficiently. Tined blades are fitted to the tillers to prepare the seedbeds quite effectively by pulvlerizing the soil. These are also used in rice fields for puddling and other operations.

(f) Earth Moving tractors - These tractors are heavy in weight and strongly built available both is tract and tyre type. These are used for earth moving work on dams, quarries and other constructional works.

Characteristics of a good tractor –

The most important requirements of a good tractor are :

1. Greater clearance, both vertical and horizontal.
2. Adaptation to the usual row-widths.
3. Quick – short turning ability.
4. Convenient and easy handling.
5. Quick and easy attachment.
6. Essential accessories such as hydraulic controls. Three point linkage and PTO.

COMPONENTS OF A TRACTOR

A tractor consists of the following main parts:

- Internal combustion engine
- Clutch
- Transmission gears
- Differential unit
- Final drive
- Rear wheels
- Front wheels
- Steering system
- Hydraulic control and hitch system
- Brakes
- Power take off unit
- Tractor pulley
- Control panel

SELECTION OF TRACTOR

The tractor should be selected based on the following factors :

1. **Land holding:** Under a single cropping pattern, it is normally recommended to consider 1 hp for every 2 hectares of land. In other words, one tractor 20-25 hp is suitable for 40 hectares farm.

2. **Cropping pattern:** Generally 1.5 hectare/hp been recommended where adequate irrigation facilities are available and more than one crop is taken. So a 30-35 hp tractor is suitable for 40 hectares farm.

3. **Soil conditions:** A tractor with less wheel base, higher ground clearance and low overall weight may work successfully in lighter soil but it will not be able to give sufficient depth in black cotton soil.

4. **Climatic conditions:** For very hot zone and desert area, air cooled engines are preferred over water cooled engines. Similarly for higher altitude, air cooled engines are preferred because water is liable to be frozen at higher altitude.

5. **Repairing facilities:** It should be ensured that the tractor to be purchased has a dealer at near by place with all the technical skills for repair and maintenance of machine.

6. **Running cost:** Tractors with less specific fuel consumption should be preferred over others so that running cost may be less.

7. **Initial cost and resale value**: While keeping the resale value in mind, the initial cost should not be very high, otherwise higher amount of interest will have to be paid.

Control Board or Dash Board of a Tractor: The control board of a tractor generally consists of the following:

- **Ignition switch:** When the ignition switch is on, the electric current flows in the electrical circuit.
- **Throttle lever:** This lever is for increasing or decreasing the speed of the engine.
- **Decompression lever:** This lever releases compression pressure from the combustion chamber of the engine and helps to stop the engine.
- **Hour meter:** This meter indicates the engine hour as well as engine revolution per minute.
- **Light switch:** Light switch is for light points only.
- **Horn button:** This is for horn of the tractor.
- **Battery charging indicator:** This indicates the charge and discharge of the battery.
- **Oil pressure indicator:** This indicates the lubricating oil pressure in the system.
- **Water temperature gauge:** This indicates the temperature of water of the cooling system.

Precautions to be taken before starting of a tractor

(1) Fuel oil should be checked in the fuel tank. If it is not adequate, fuel oil should be added to the tank.

(2) Lubricating oil should be checked by a dipstick and if necessary it should be topped up.

(3) Water in the radiator should be checked, if necessary it should be topped up.

(4) Air cleaner should be checked to see whether it is clean or blocked. If blocked, it should be cleaned.

(5) Transmission oil should be checked by a dipstick, if necessary it should be topped up.

(6) Air pressure in the tyres should be checked and if necessary the tyres should be inflated as recommended by the manufacturers.

(7) Fan belt should be checked by hand; if necessary it may be tightened or loosened.

(8) Grease points should be checked, to see whether they have been greased or not.

(9) Important nuts and bolts should be checked. If any of them are loose, it should be tightened.

(10) The water level of the battery should be checked. If it is below the partition wall, it should be filled up with distilled water.

Method of starting a diesel tractor

In order to start a tractor, the following sequence should be followed:

(1) Open the fuel cock.

(2) Put the gear shift lever and PTO lever into neutral position.

(3) Put the throttle lever in about 3/4th position.

(4) Put the hydraulic control level to the lowered position.

(5) Press the clutch pedal and turn the ignition key to the on position. Thus the tractor will be started.

Method to stop a tractor

Tractor is a heavy machine and it can cause serious accident if it is not stopped in time. The following procedure should be followed for stopping the movement of the tractor.

(1) Move the throttle lever down and reduce the engine speed to the lowest possible limit.

(2) Press the clutch pedal and press the brake pedal of the tractor to stop the motion of the tractor.

(3) Put the gear shift lever into neutral position.

(4) If an implement is attached to the tractor, hydraulic control lever should be moved slowly to the lower position.

(5) Pull the stock lever.

(6) Apply the parking brakes, if necessary.

Outlets of a tractor : There are four power outlets of tractor :

POWER TAKE-OFF UNIT (PTO)

It is a part of tractor transmission system. It consists of a shaft, a shield and a cover. The shaft is externally splined to transmit torsional power to another machine as shown in Plate : 10.1. A rigid guard fitted on a tractor covers the power take-off shaft as a safety device. This guard is called power take-off shield. Agricultural machines are coupled with this shaft at the rear part of the tractor. As per ASABE standard pto speed is 540±10 rpm when operating under load. In order to operate 1000 rpm drive machine, a new standard has been developed. Modern tractors are provided with pto speed of 1000±10 rpm.

Plate : 10.1 Tractor operator PTO shaft and it's operation

Hook

A hook (Plate : 10.2) is provided for hitching of trailers, and other stationery machines operated by tractor like thresher, winnower etc. It is provided at the back of the tractor.

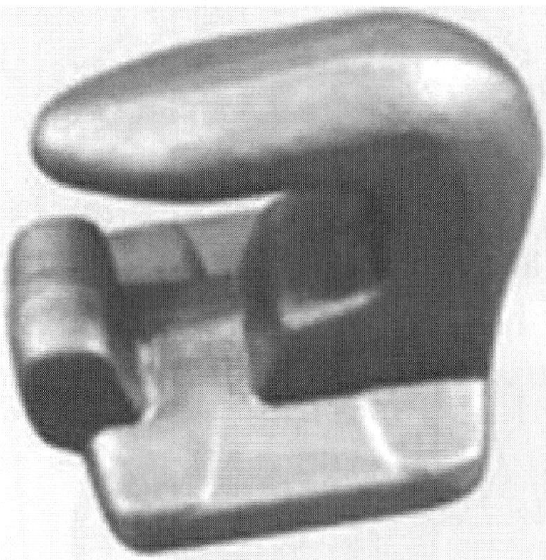

Fig : 10.2 Tractor hook

(Courtesy: http://daimlersteelforge.com) Three point linkage

It is a combination of three links, one is upper link and two are lower links, the links articulated to the tractor and the implements at their ends in order to connect the implement to the tractor. All the tractor drawn implement are attached to this and hydraulic control is also provided for three point linkage. The Three point linkage is shown in Plate : 9.3.

Fig : 10.3 There point linkage of tractor
(Courtesy: http://www.agmax.com.au)

Belt pulley

A pulley is provided with pto shaft for some of the tractors through which rotary power output can be tapped for operating stationery machines.

Power tiller

The power tiller (Plate : 10.4) is a walking type tractor in which the direction of travel and its control for field operation is performed by the operator walking behind it. It is also known as Hand tractor or Walking type tractor. The concept of also known as Hand tractor or walking tractor. The operator walks behind the power tiller, holding the two handles of power tiller in his hands. Power tiller may be called a single axle walking type tractor, though a riding seat is provided in certain designs. The concept of power tiller came in the world in the year 1920. Japan is the first country to use power tiller was designed in the year 1947. Production of power tiller rapidly increased during the year 1950 to 1965. Power tiller was first introduced in India in the year 1963. Manufacturing of several makes of power tillers like Iseki, Sato, Krishi, Kubota, Yanmar and Mitsubishi were started in India after 1962.

Fig: 10.4 Power tiller

(Courtesy: http://www.china-tractors.com) Components of power tiller

A power tiller consists of the following main parts:

1. Engine
2. Transmission gears
3. Clutch
4. Brakes
5. Rotary unit

All the power tillers are fitted with an I.C. engine. At present, most of the power tillers are fitted with diesel engine. Only Iseki make have used kerosene engine. Other makes like Kamco, Mitsubishi, Krishi, Yanmar and Satoh have used diesel engine in India.

OPERATION

The main clutch is a lever on the handle. The lever can be shifted to on or off position while operating in the field. When the lever is shifted to on position, the power from the engine is transmitted through the main clutch to the various parts of power tiller. When the lever is shifted to off position, the power from the engine is cut-off from the rest of the transmission.

Power transmission in power tiller

For operation of power tiller, the power is obtained from the I.C. engine, fitted on the power tiller. The engine power goes to the main clutch with the help of belt or chain. From main clutch, the power is divided in two routes, one goes to transmission gears, steering clutch and then to the wheel. The other component goes to the tilling clutch and then to the tilling attachment.

Components of a power tiller

No.	Name of Parts	No.	Name of Parts
1.	Handle grip	21.	V – belt
2.	Throttle lever	22.	Tension pulley
3.	Auxiliary handle	23.	Main clutch pulley
4.	Main gearshift lever	24.	Belt cover
5.	Hanger	25.	Handle cover
6.	Front stand operating lever	26.	Handle cover grip
7.	Auxiliary chain case set screw	27.	Tine speed change lever
8.	Auxiliary chain case	28.	Main gear shift lever
9.	Steering clutch lever (right)	29.	Hand light
10.	Rear wheel height adjusting lever	30.	Lamp switch
11.	Rear wheel pipe attaching handle	31.	Steering clutch lever (left)
12.	Ridger set screw	32.	Steering clutch wire
13.	Mud – guard (right)	33.	Rubber guard
14.	Side cover (right)	34.	Protector
15.	Tilling tines	35.	Fender left
16.	"Magic Bar"	36.	Rubber tire
17.	Front frame	37.	Hexagon wheel the
18.	Protector	38.	Side Frame
19.	Front lights	39.	Tilling tines
20.	Engine pulley	40.	Rear wheel

The flow diagram for transmission of power is given below:

Engine Main clutch

Transmission gear Steering clutch Wheels

Tilling clutch Tilling attachment

V-belt is usually used to transmit power from the engine to the main clutch, because V – belt has very high efficiency and it works as a shock absorber also.

Engine : Single cylinder, diesel, air/water cooled, 8-15HP.

Main clutch

Power goes from the engine to the main clutch. Clutch may be:

(a) Friction clutch or

(b) V – belt tension clutch

Friction clutch is generally used for bigger power tiller. Usually it is a dry type multiple disc clutch. V – belt tension clutch is used for small power tillers. The main function of clutch in a power tiller are:

(i) To transmit engine power to transmission gears and

(ii) To make power transmission gradually and smooth

Trouble :

The clutch slips continuously, the causes and remedies are :

S.No	Causes	Remedy
1	Oil inside the clutch	Clean with gasoline and dry it
2	Worn lining	Replace the lining
3	Spring tension not sufficient	Replace the spring
4	Water inside the clutch	Dry it with heater etc.
5	Poor adjustment of clutch	Adjust the clutch
6	Poor contact of friction surface	Grind the surface

Transmission gears : Transmission box consists of gears, shafts and bearings. The speed change device may be either Gear type or belt type.

Brakes : All power tillers have some braking arrangement for stopping the movement. Most of the power tillers use inner side expansion type brake.

Wheels : Usually 2 pneumatic tyres are used in power tillers. The pressure of the type ranges from 1.1 to 1.4 kg/cm2.

Rotary unit : Power tiller has a rotary unit for field operation which may be centre drive type or side drive type. The power tiller operated rotary tiller is shown in Plate : 10.5. The centre drive type has got transmission at the centre and the side drive type the transmission at one side.

Fig : 10.5 Power tiller operated tiller

(Courtesy: http://web.tradekorea.com)

The centre drive type has the following characteristics:

(a) Tilling width can be widened

(b) Rotary unit is light in weight

(c) Fixing of attachment is easy

(d) The tine shaft can be detached easily

(e) Mounting and dismounting of rotary unit is very easy

(f) It may leave some portion of the field untilled

(g) It has one point support on the ground

The side drive type has the following characteristics:

(a) Deeper tilling is possible

(b) The arrangement is useful for hard soil

(c) It has two point support on the ground

Rotary tines: Rotary tines are used in rotary unit for soil cutting and pulverization purpose. Rotary tines are of three types namely, straight tines, curved tines and sliding tines.

Steering clutch lever: Steering clutch is provided on the grip of the right and left handles. When the left side is gripped, power is cut-off on left side of the wheel and the power tiller turns to the left. Similarly when the right side is gripped, the power tiller turns to the right.

11

ELECTRIC MOTORS

Electric motor is a machine which converts electrical energy in to mechanical energy at much higher efficiencies than I.C. engines. Electrical motors operate at efficiencies ranging from 50 to 90 % as against I.C. engines having thermal efficiencies between 28 to 38 %. The most important feature are the low initial cost, low operating cost, easy to start even at loads etc. besides, the motors are designed to take up temporary over loads. These motors have long life, compact construction, low noise levels, nothing like exhaust smokes, operational simplicity and minimum safety hazards.

A motor has primarily consists of armature coil and a field coil. When armature coil are supplied with electric current from the supply and the field magnets are excited and they experience a force tending to rotate the armature. Each conductor experiences a force which tends to rotate the armature in either clockwise or anticlockwise direction. These forces collectively produce a twisting effect or a torque. This causes the armature to rotate in one of the direction.

The electric motors are classified as Direct current motor (D.C.Motor) and Alternating current motor (A.C.Motor). Electric motors are also classified on the basis of armature designs, rotor design and kilo watt power output.

D.C. MOTOR

An electric motor is a machine which when supplied with electric current, can be used for some mechanical work. Direct current motors are widely used in applications that require control of speed e.g., electric trams, engine starters and elevators etc. They are made in sizes from 1/100 hp to thousands of hp. D.C. motors are similar to D.C. generators in appearance and construction. They have a similar armature, field poles and other motors, current is conducted to the armature winding. This is accomplished by connecting leads from the armature winding to the commutator, and then feeding the commutator with current. The commutator is supplied with current by means of carbon brushes. There are three main types of D.C. motors: the series motor, the shunt motor and the compound motor. These motors differ in the construction of their field coils and in the connections between the field coils and armature.

Small D.C. motors of less than 1/2 kW size consume very little current and therefore, can be started by placing full voltage across the motor terminals. But in large D.C. motors, if full voltage is applied, excessive current will flow which damages the motor. To start a large motor, it is necessary to connect a resistance unit in series with the armature, so that the starting current is reduced to a safe value. As the motor starts, the resistance is completely cut off after the motor has reached the desired speed. At high speed the counter emf produced in the armature prevents an excessive current flow. Rotation of D.C. motors can be changed by reversing the direction of the current flow in either the armature or field circuit, but not both. Generally, it is reversed by changing the direction of current flow in the armature circuit.

D.C.Motor Starter : The operation of the starter of a D.C. motor is usually done manually. When the handle is brought to make the current with the first point, the shunt field is immediately connected directly across the supply line. The contact with the point first the shunt field is immediately connected directly across the supply line. The contact with the point also connects the entire resistance in series with the armature which begins to rotate and the initial current is limited to a safe value. As the handle is moved current up, the resistance is gradually removed from the armature circuit. When all the resistance is finally removed, the operating handle rests against a holding magnet mounted along side the last point. The holding magnet is connected in series with the main-shunt field of the motor. The operating handle is provided with a heavy spring which will instantly return it to the off position if the holding magnet is demagnetized due to failure of the power supply or break in the motor field circuit. In such a case the handle automatically returns to the of position and the motor stops. It will remain so until started again by hand.

A.C. MOTORS

For most farm work where electric motors can be used, induction motors are generally preferred. Induction motors may be classified as :

1. **Single phase motors :** This type of motors are further classified as (a) split phase type, (b) Capacitor start type, (c) Shaded pole type, (d) Wound rotor type, (e) Synchronous type (f) Universal type.

2. **Poly phase motors:** two types : (a) Two phase motors and (b) Three phase motors.

Among the single phase motors, the first two types are very much in use and they are described in detail here.

Split-Phase Motor: This motor has four main parts, namely rotor, stator, end plates, and centrifugal switch. It is generally operated from a single phase lighting or power circuit. The rotor consists of three parts, the laminated core, the shaft and the squirrel cage winding. The third part consists of heavy copper bars which are placed in slots in iron core and which connected to each other by means of heavy copper rings placed on both ends of the core.

The stator is composed of a laminated iron core with semi closed slots, a heavy cast iron frame, and two windings of insulated copper wires that are wound into the slots and are called the running and starting windings. Both the winding are connected to the power line when the motor is started. As the current starts flowing through both the windings, it causes a magnetic field to be formed inside the motor. This magnetic field rotates and induces a current in the rotor winding which in term causes another magnetic field. These combine in such a manner as to causes rotation of the rotor. No sooner does the motor reach approximately 75 per cent of the full speed and then the starting windings is cut out of the circuit by means of centrifugal switch.

The end plates are fastened to the stator frame and mainly keep the rotor in position. End plates are fitted with either ball bearings or sleeve bearings in which the rotor shaft rotates.

Capacitor Motor: The construction of the capacitor motor is similar to that of the split phase motor. But an additional part known as the 'Capacitor' is also provided with these machines. The capacitor is usually mounted on the top of the motor. It is connected in series with the starting winding and, therefore, higher starting torque with lower starting current is developed. A capacitor has capacity to store electricity. It is formed when two conductors are separated by an insulator, such as waxed paper. Capacitors are rolled and placed in a metal container. They are rated in microfarads. Depending upon the size of the motor, its capacity may vary from 10 to 150 microfarads.

To produce a starting torque in a capacitor motor, a revolving magnetic field must be established inside the motor. This is accomplished by placing the starting winding 90 electrical degrees out of phase with the running winding. The capacitor helps the current in the starting winding to reach the maximum value before the current in the running winding becomes a maximum. This condition produces a revolving magnetic field in the stator which in turn induces current in the rotor winding. As a result the magnetic field acts in such a manner as to produce rotation of the motor.

Three Phase Induction Motor: Three Phase Induction Motor has three main parts: stator, rotor and end plates. Its construction is very similar to that of the split phase motor, but it is not provided with centrifugal switch. The stator consists of cast iron frame and laminated iron core.

The coils in the slots of the iron core are connected to form three separate windings called phases. These windings are connected so that a magnetic field is foamed inside the stator and that causes the rotor to turn at a constant speed. The squirrel cage type rotor is mostly used. The end plates are meant for supporting the rotor shaft. A.C.Motor

Starters: If an A.C. motor is started on full line voltage, it will draw about four times its normal running current. On very large motors, it is always desirable to reduce the starting current, otherwise damage may be done to the motor and line disturbances may be created affecting operation of other motors, on the same line. For single phase small motors (below 1 kW), a hand operated switch is used to control the motor. In the case of medium size motors (2hp or more), it is necessary to insert in the line some device which will reduce the starting current. Push button type starters or direct on line starters and star-delta-starters are widely used for three phase up to 7.5 kW motors above 7.5 kW only star-delta starter is used.

Push Button Starter : This consists of two push buttons, one for starting (usually green) and other for stopping (usually red) the motor. Pressing the green button causes the contacts inside the switch to make and connect the motor across the line. Pressing the red or stop button causes the contacts to break apart and open the circuit to the motor. Generally the push button starter is equipped with an overload safety device. It opens the circuit to the motor if an over load current flows for a short period of time.

Star-delta-Starters : This starter uses the reduced voltage starting principle only to three phase- delta connected motors. While starting, the motor is connected to star and thus the line voltage is reduced to $1/\sqrt{3}$ value which each phase of the motor received then (400 line voltage reduces to 230 volts). The reduced phase voltage thus reduces the phase current in the same proportion. To make use of this type of starter, it is necessary to bring 6 leads out of the motor so that they can be changed from star at starting time in delta in the running condition. In starting, as the starting lever is pressed down, the motor starts rotating as a star connected motor. After the motor accelerates the starting lever is raised up, thus the switch is closed in the running position. The motor now runs on full voltage.

SELECTION OF ELECTRIC MOTORS

The starting requirements of equipment and the maximum current that may be drawn from the single phase power service are the two most important requirements for selecting an appropriate motor. However, choosing a motor that will meet load requirements without exceeding temperature and torque limits are also equally important considerations. Hence, the first step in motor selection is to determine the load characteristic such as power or torque requirements, speed and duty cycle.

To select the right type of motor for any particular job, the following factors should be considered:

1. The characteristic of the power supply whether A.C or D.C. ; single or three phase, 50 or 60 cycles supply is available.
2. Horse Power requirement - to avoid over loading, slightly higher size should be preferred.

3. Duty requirement - continuous or intermittent duty.
4. Speed of operation - should be selected with the consideration of the machine to be driven.
5. Starting current requirement should be known before selecting the motor for a particular use.
6. Atmospheric and environmental conditions-humidity, splashing of liquids, excess dust etc. should be kept in view.
7. Over load protection device-this always desirable if the motor is equipped with such a device.
8. Price-it is quite logical to select the lowest price motor that will do the job satisfactorily.
9. However, the price should not be the first consideration.

By measuring the mechanical output of a motor and the electrical input to the motor under different load conditions with the help of 'Prony Brake' dynamometer, the torque-speed characteristic curves can be developed.

CARE OF ELECTRIC MOTORS

1. Keep the motor bearings properly lubricated.
2. Keep the motor clean and well ventilated.
3. Never overload the motor and thus avoid damage to the motor due to burning of the windings.
4. Keep the belt moderately tight between motor and driven machine.
5. Never connect the motor to too low or high supply voltage. Both conditions are harmful.
6. Keep the motor grounded.
7. Keep the motor away from moisture, dust and saw dust.

How to Clean a Motor

Electric motors must be kept clean and lubricated if they are expected to work efficiently for a longer period. The following procedure should be observed to keep a motor clean :

1. After having disconnected the motor from the power, it is necessary to disassemble it by removing the rotor, shaft and fan assembly. All possible care should be taken not the break any wire connection or centrifugal switch mounted on the rotor shaft or elsewhere.

2. The second step is to blow out the loose dirt from the motor windings and from inside the end plates by using pressurized air from a blower or from a cycle pump. If necessary, a dry paint brush or cloth can be used to carefully wipe off dirt that might remain even after using pressurized air.
3. All dirt from the air passages in the frame and in the rotor should be removed. Metal parts may be washed with non- inflammable fluid.
4. All the accessories like the starting switch, the commutator and brushes should be checked and repaired before the parts are reassembled.
5. Lastly the motor bearings are lubricated with the recommended type of lubricant before the connection is provided to check its operation.

Fuse: Fuse generally means a fuse wire, placed in a fuse holder. It is a safety device, which protects electrical/electronic circuits against over loads, short circuits and earth faults. The fuse is connected in the phase of the supply. When the current taken by the circuit exceeds the rated current of the fuse wire, the fuse wire melts and cuts. This disconnects the supply from the circuit and thus protects the circuit and the components connected in the circuit. Rating of the fuse: The maximum current that a fuse can carry, without being burnt, is called the rating of the fuse wire. It is expressed in Amperes.
Ex. 5 A fuse, 10 A fuse, 16A fuse, 60A fuse, 100 A fuse, etc. Current rating of the fuse, selected for a circuit, should be equal to the maximum current rating of the machinery, appliance or component connected in the circuit.

12

TILLAGE

It is a mechanical manipulation of soil to provide favorable condition for crop production. Soil tillage consists of breaking the compact surface of earth to a certain depth and to loosen the soil mass, so as to enable the roots of the crops to penetrate and spread into the soil.

OBJECTIVE OF TILLAGE

- To obtain seed bed, suitable for different type of crops.
- To add more humus and fertility to soil by covering the vegetation.
- To destroy and prevent weeds.
- To aerate the soil for proper growth of crops.
- To increase water absorbing capacity of the soil.
- To destroy the insects, pests and their breeding places and
- To reduce the soil erosion.

Classification of Tillage

Tillage is divided into two classes: i) Primary tillage, ii) Secondary tillage.

Primary tillage

It constitutes the initial major soil working operation. It is normally designed to reduce soil strength, cover plant materials and rearrange aggregates. The operations performed to open up any cultivable land with a view to prepare a seed bed for growing crops is known as primary tillage. Implements may be tractor drawn or animal drawn implements. Animal drawn implements mostly include indigenous plough and mould-board plough. Tractor drawn implements include mould-board plough, disc plough, subsoil plough, chisel plough and other similar implements.

Secondary tillage

Tillage operations following primary tillage which are performed to create proper soil tilth for seeding and planting are secondary tillage.

These are lighter and finer operations, performed on the soil after primary tillage operations. Secondary tillage consists of conditioning the soil to meet the different tillage objectives of the farm. The implements used for secondary tillage operations are called secondary tillage implements. They include different types of harrow, cultivators, levellers, clod crushers and similar implements. These operations are generally done on the surface soil of the farm. Secondary tillage operations do not cause much soil inversion and shifting of soil from one place to other. These operations consume less power per unit area compared to primary tillage operations.

- To break the big clods and to make the field surface uniform and levelled.
- To cut crop residues and mix them with top soil of the field.

Secondary tillage implements may be tractor drawn or bullock drawn implements.

Bullock drawn implements include harrows, cultivators, hoes etc.

Terms Used in tillage

Tool - It is an individual working element such as disc or shovel.

Implement - It is an equipment generally having no driven moving parts, such as harrow or having only simple mechanism such as plough.

Machine -It is a combination of rigid or resistant bodies having definite motions and capable of performing useful work.

Plough

The main implement for primary tillage is plough, which is used for ploughing operations. Ploughing is the primary tillage operations, which is performed to cut, break and invert the soil partially or completely. Ploughing essentially means opening the upper crust of the soil, breaking the clods and making the soil suitable for sowing seeds. The purpose of ploughing is to achieve the objectives of the tillage.

Types of ploughing

Normal Ploughing : It is the ploughing up to a depth of about 15 cm.

Contour Ploughing: It is the method of ploughing in which the soil is broken and turned along the contours.

PLOUGHING OF LAND

The ploughing of land is to separate the top layer of soil into furrow slices. The furrows are turned sideways and inverted to a varying degree, depending upon the type of plough being used. It is a primary tillage operation, which is performed to shatter soil uniformly with partial or complete soil inversion. There are a few important terms frequently used in connection with ploughing of land.

(i) Furrow - It is a trench formed by an implement in the soil during the field operation.

(ii) Furrow slice- The mass of soil cut, lifted and thrown to one side is called furrow slice.

(iii)Furrow wall - It is an undisturbed soil surface by the side of a furrow.

(iv) Crown- The top portion of the turned furrow slice is called crown.

(v) Back furrow - A raised ridge left at the centre of the strip of land when ploughing is started from centre to side is called back furrow. When the ploughing is started in the middle of a field, furrow is collected across the field and while returning trip another furrow slice is lapped over the first furrow. This is the raised ridge which is named as back furrow.

(vi) Dead furrow - An open trench left in between two adjacent strips of land after finishing the ploughing is called dead furrow.

(vii) Head land - While ploughing with a tractor to turn, a strip of unploughed land is left at each end of the field for the tractor to turn, that is called head land. At the end of each trop, the plough is lifted until the tractor and the plough have turned and are in position to start the return trip. The head land is about 6 metres for two or three bottom tractor plough and one metre more for each additional furrow. The Line diagram of different terminology used in ploughing is shown in Fig.12.1.

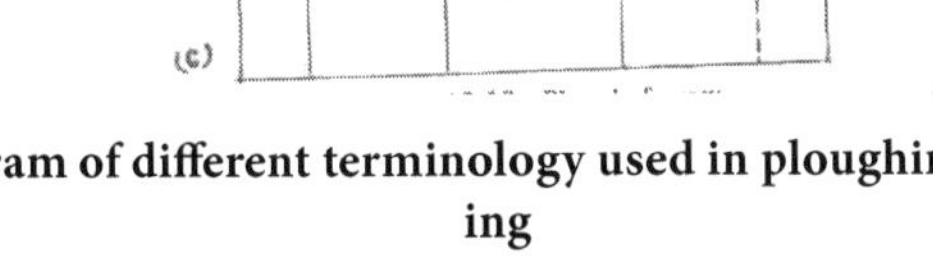

Fig. 12.1 Line diagram of different terminology used in ploughing Methods of ploughing

In order to provide furrows at all times on the right hand side of the plough, two method of working are used called Gathering and Castering (Fig.12.2).

Gathering - Whenever a plough works round a strip of ploughed land, it is said to be gathering. Casting- Whenever a plough works round a strip of unploughed land, it is said to be casting. Ploughing of a field by casting or gathering alone is normally uneconomical. The following are a few important methods used in tractor ploughing.

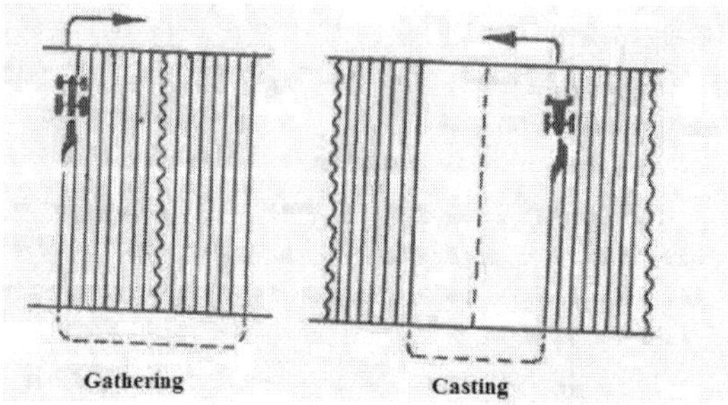

Fig. 12.2 Line diagram if plouging methods Continuous ploughing method

In normal conditions, the continuous ploughing method (Fig. 12.3) is considered very convenient and economical. This is a method usually used in which the tractor and plough never run idle for more than three quarter land width along the headland and never turn in a space narrower than a quarter land width. In this method, first the headland is marked and the first ridge is set up at three quarter of a land width from the side. The other ridges are set at full width over the field. The operator starts ploughing between the first ridge and the side land. The operator continues to turn left and cast in the three quarter land until a quarter land width of ploughing is completer on each side. At this stage, it is important to lift the plough to half depth for last trip down the side land of the field. This leaves a shallow furrow where the finish comes.

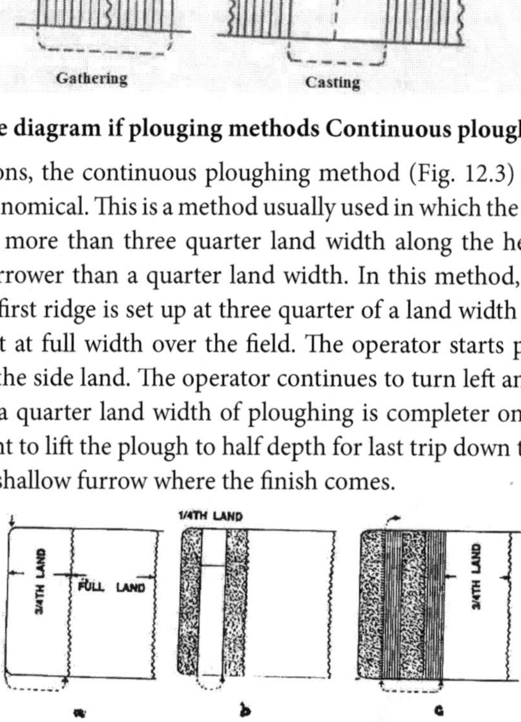

Fig. 12.3 Line diagram of Continuous ploughing method

After this, the driver turns right and gathers round the land already ploughed on the first ridge. Gathering is continued till the unploughed strip in first three-quarter land has been ploughed and completed. This gathering reduces the first full land by a quarter. The remaining three quarter land can be treated in exactly the same manner as the original three quarter land. This process is repeated for all other lands in the field.

Round and round ploughing

In this method, the plough moves round and round a field. This system is adopted under conditions where ridges and furrows interfere with cultivation work. The field can be started in two ways.

(a) Starting at the centre

A small plot of land is marked in the middle of the field and it is ploughed first. After that, the plough works round this small plot and the entire plot is completed. This is not a very economical method.

(b) Starting at the outer end

Tractor starts ploughing at one end of the field and then moves on all the sides of the plot and comes gradually from the sides to the centre of the field. Wide diagonals are left unploughed to avoid turning with the plough. There are no back furrows in this method. Conventional ploughing is usually done by this method.

One way ploughing

This system requires the use of a special type of plough known as reversible plough or one way plough. Such a plough turns furrows to the left or right. After the headland has been marked, the operator plough along a straight side land mark. At the end of the first trip, he turns his tractor in a loop and returns down the same furrow. No dead and back furrows are left in the field. In gently sloping fields, this method is suitable.

13

PRIMARY TILLAGE IMPLEMENTS

INDIGENOUS PLOUGH

Indigenous plough is most commonly used in this country as a primary tillage implement. The shape and size of the plough varies with places and regions due to variation in soil types and tillage requirements. The bullock drawn indigenous plough is shown in Plate : 13.1. The main parts of the plough are: body, share, shoe, beam and handle. The body is the main frame to which the shoe, beam and handle are attached. The share is a narrow steel bar attached to the upper surface of the shoe longitudinally along the centre line and the handle is generally attached to the body of the plough. The share is attached to the shoe which penetrates into the soil and breaks it open. The shoe also helps in stabilizing and balancing the plough while in operation. The plough is provided with a wooden beam and a handle. The indigenous plough is made of wood except the share which is made of steel bar.

Fig : 13.1 Bullock drawn indigenous plough Mould board plough

A mould board plough is very common implement used for primary tillage operations and is shown in fig. : 13.2.

Fig : 13.2 Two bottom fixed Mould board plough

The main functions of mould board plough are:

- Cutting the furrow slice
- Lifting the soil
- Turning the furrow slice
- Pulverizing the soil.

Components of mould board plough

A mould board plough consists of the different parts like Share, Mould board, Land side, Frog and Tail piece. The different parts of mould board plough are shown in Fig. 13.1

1. **Share:** It is that part of the plough bottom which penetrates into the soil and makes a horizontal cut below the surface.
2. **Mould board** - It is the curved part which lifts and turns the furrow slice.
3. **Land side** - It is the flat plate which bears against and transmits the rear side lateral thrust of the plough bottom to the furrow wall.
4. **Frog** - It is the part to which other components of the plough bottom are attached.
5. **Tail piece** - It is an adjustable extension, which can be fastened to the rear of a mould board to help in turning a furrow slice.

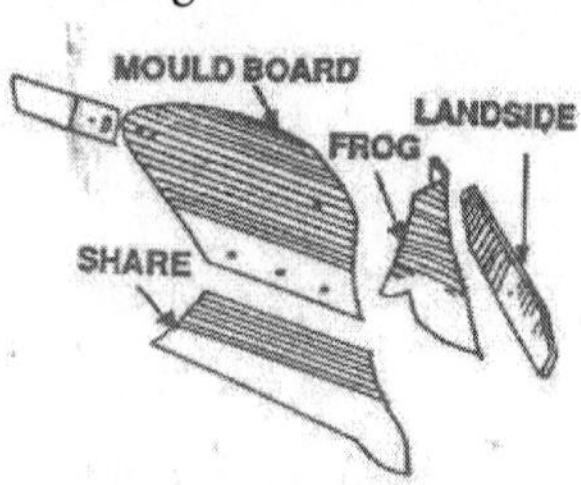

Fig. 13.3 Components of Mould Board plough

The share is one of the important part of the plough, which penetrates into the soil and makes a horizontal cut below the soil surface. It is a sharp, well polished and pointed component. The different parts of the share are shown in Fig. 13.2.

1. Share point : It is the forward end of the cutting edge which actually penetrates into the soil .

2. Cutting edge: It is the front edge of the share which makes horizontal cut in the soil.

3. Wing of share: It is the outer end of the cutting edge of the share. It supports the plough bottom .

4. Gunnel: It is the vertical face of the share which slides along the furrow wall. It takes the side thrust of the soil and supports the plough bottom against the furrow wall.

5. Cleavage edge: It is the edge of the share which forms joint between mould board and share on the frog.

6. Wing bearing: It is the level portion of the wing of the share, providing a bearing for the outer corner of the plough bottom.

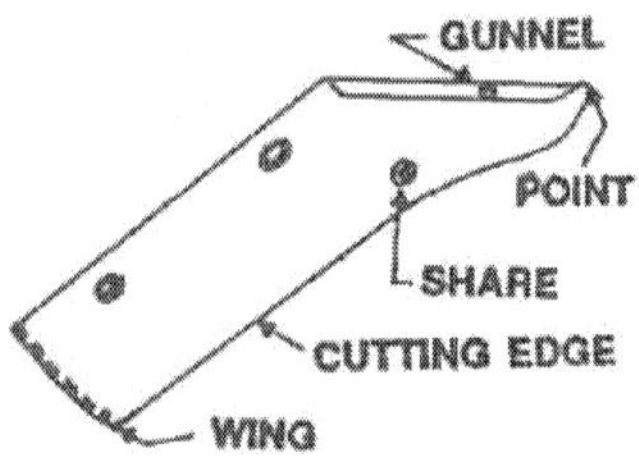

Fig. 13.4 Different parts of the share

Material of share

The shares are made of chilled cast iron or steel. The steel mainly contains about 0.70 to 0.80% carbon and about 0.50 to 0.80% manganese besides other minor elements.

Type of Share

There are four different types of shares used for the mould board plough such as Slip share, Slip nose share, Shin share and Bar point share and are shown in Fig.13.3.

i) Slip share

It is one piece share with curved cutting edge, having no additional part. It is a common type of share, mostly used by the farmers. It is simple in design, but it has got the disadvantage that the entire share has to be replaced if it is worn out due to constant use.

ii) Slip nose share

It is a share in which the point of share is provided by a small detachable piece. It has the advantage that share point can be replaced as and when required. If the point is worn out, it can be changed without replacing the entire share, effecting considerable economy.

iii) Shin share

It is the share having a shin as an additional part. It is similar to the slip share with the difference that an extension is provided to it by the side of the mould board.

iv) Bar point share

It is the share in which the point of the share is provided by an adjustable and replaceable bar. This bar serves the purpose of point of the share and land side of the plough.

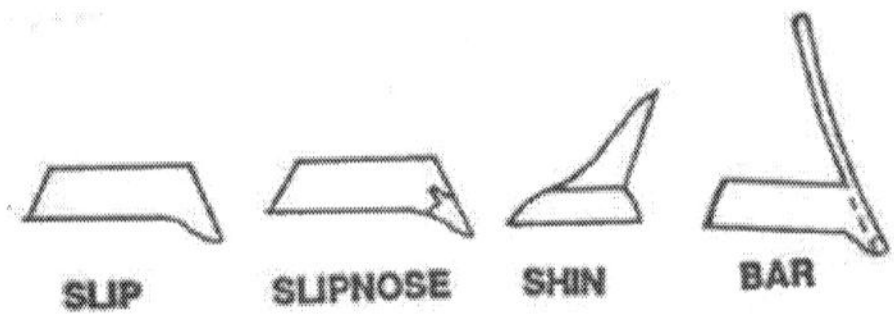

Fig. 13.5 Different types of share

2. Mould board

The mould board is that part of the plough which receives the furrow slice from the share. If lifts, turns and breaks the furrow slice. To suit different soil conditions and crop requirements, mould board has been designed in different shapes. There are four types of mould board used such as : General purpose, Stubble, Sod or Breaker and Slat.

GENERAL PURPOSE

It is a mould board having medium curvature lying between stubble and sod. The sloping of the surface is gradual. It turns the well defined furrow slice and pulverizes the soil thoroughly. It has a fairly long mould board with a gradual twist, the surface being slightly convex. The general purpose mould board is shown in Fig.13.4

Fig. 13.6 General purpose mould board

Stubble type

It is short but broader mould board with a relatively abrupt curvature which lifts breaks and turns the furrow slice used in stubble soils. Its curvature is not gradual but it is abrupt along the top edge. This causes the furrow slice to be thrown off quickly, pulverizing it much better than other types of mould board. This is best suited to work in stubble soil that is under cultivation for years together. Stubble soil is that soil in which stubble of the plants from the previous crop is still left on the land at the time of ploughing. This type of mould board is not suitable for lands full of grasses and is shown in Fig. 13.5.

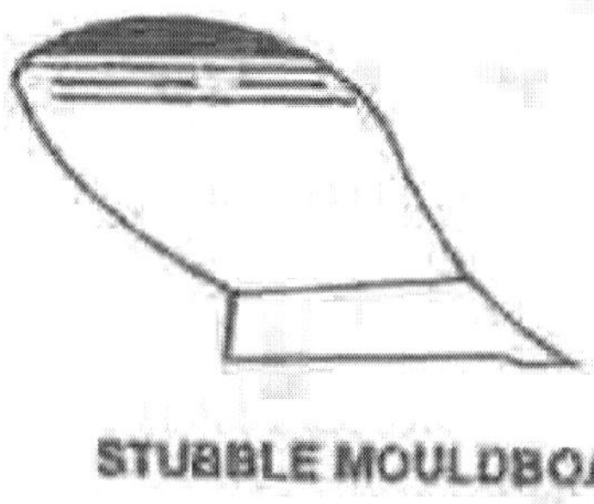

Fig. 13.7 Stubble mould board

Sod or Breaker type

It is a long mould board with gentle curvature which lifts and inverts the unbroken furrow slice. It is used in tough soil of grasses. It turns over thickly covered vegetable of the soil. This is very useful where complete inversion of soil is required by the farmer. This type has been designed for used in sod soils. The sod or breaker type mould board is shown in Fig. 13.6.

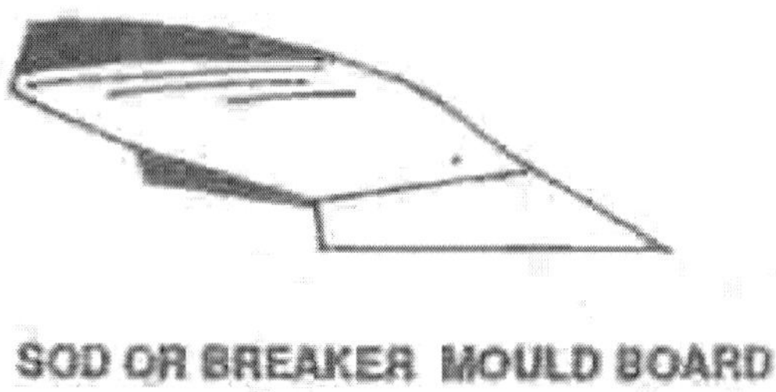

Fig. 13.8 General purpose mould board

SLAT TYPE

It is a mould board whose surface is made of slats placed along the length of the mould board, so that there are gaps between the slats. This type of mould board is often used, where the soil is sticky, because the solid mould board does not scour well in sticky soils and is shown in Fig.13.7.

Fig. 13.9 Slat mould board

3. Land side

It is the flat plate which bears against and transmits lateral thrust of the plough bottom to the furrow wall. It helps to resist the side pressure exerted by the furrow slice on the mould board. It also helps in stabilizing the plough while it is in operation. Land side is fastened to the frog with the help of plough bolts. The rear bottom end of the land side is known as heel which rubs against the furrow sole.

4. Frog

Frog is that part of the plough bottom to which the other components of the plough bottom are attached (Fig.6). It is an irregular piece of metal. It may be made of cast iron for cast iron ploughs or it may be welded steel for steel ploughs.

5. Tail piece

It is an important extension of mould board which helps in turning a furrow slice.

PLOUGH ACCESSORIES

There are a few accessories, which are used to increase the efficiency of ploughing. The different accessories used are Jointer, Coulter, Gauge wheel, Land wheel and Furrow wheel.

1. Jointer

It is a small irregular piece of metal having a shape similar to an ordinary plough bottom. It looks like a miniature plough. Its purpose is to turn over a small ribbon like furrow slice directly in front of the main plough bottom. This small furrow slice is cut from the left and upper side of the main furrow slice and is inverted so that all trashes on the top of the soil are completely turned down and buried under the right hand corner of the furrow.

2. Coulter

It is a device used to cut the furrow slice vertically from the land ahead of the plough bottom. It cuts the furrow slice from the land and leaves a clear wall. It also cuts trashes which are covered under the soil by the plough. The coulter may be (a) Rolling type disc coulter and (b) Sliding type knife coulter.

a. Rolling type disc coulter

It is a round steel disk which has been sharpened on the edge and suspended on a shank and yoke from the beam. The edge of the coulter may be either smooth or notched. It is so fitted that it can be adjusted up-down and side ways. The up-down adjustment takes care of depth and sideways adjustment is meant for taking care of width of cut.

b. Sliding type knife coulter

It is a stationary knife fixed downward in a vertical position on the beam. The knife does not roll over the ground but slides on the ground. The knife may be of different shapes and sizes.

3. Gauge wheel

It is an auxiliary wheel of an implement to maintain an uniform depth of working. Gauge wheel helps to maintain uniformity in respect of depth of ploughing in different soil conditions. It is usually placed in hanging position.

4. **Land wheel** - It is the wheel of the plough which runs on the unploughed land.

5. **Furrow wheel** – There are two types of furrow wheel.

Front furrow wheel - It is the front wheel of the plough which runs in the furrow. Rear furrow wheel - It is the rear wheel of the plough which runs in the furrow. Adjustment of mould board plough

For proper penetration and efficient work by the mould board plough, some clearance is provided in the plough. This clearance is called suction of the plough. Suction in mould board plough is of two types (i) Vertical suction and (ii) Horizontal suction.

VERTICAL SUCTION (VERTICAL CLEARANCE)

It is the maximum clearance under the land side and the horizontal surface when the plough is resting on a horizontal surface in the working position. It is the vertical distance from the ground, measured at the joining point of share and land side. It helps the plough to penetrate into the soil to a proper depth. This clearance varies according to the size of the plough. The vertical clearance is shown in Fig.13.8.

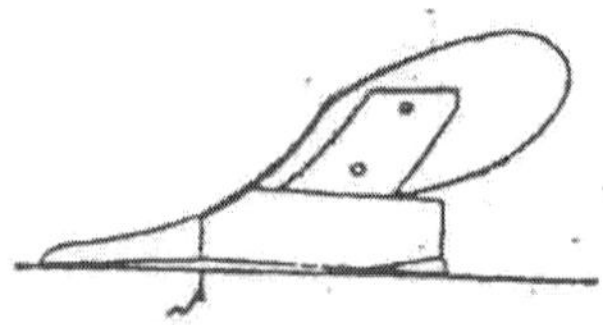

Fig. 13.10 Vertical suction of mould board

Horizontal suction (Horizontal clearance)

It is the maximum clearance between the land side and a horizontal plant touching point of share at its gunnel side and heal of land side. This suction helps the plough to cut the proper width of furrow slice. This clearance varies according to the size of the plough. It is also known as side clearance and is shown in Fig.13.9.

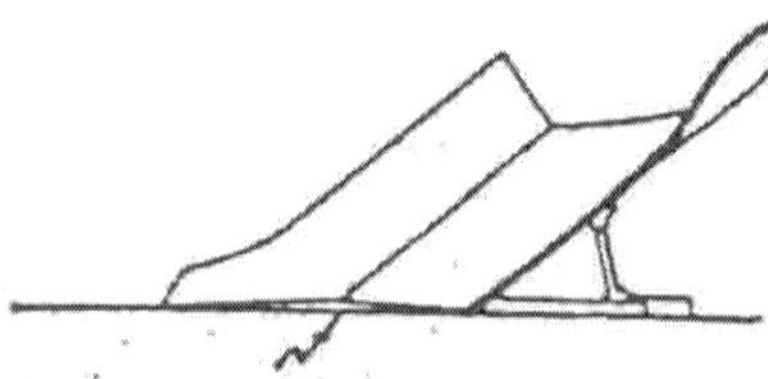

Fig. 13.11 Horizontal suction of mould board

Throat clearance

It is the perpendicular distance between point of share and lower position of the beam of the plough. The throat clearance of mould board plough is shown in Fig. 13.10.

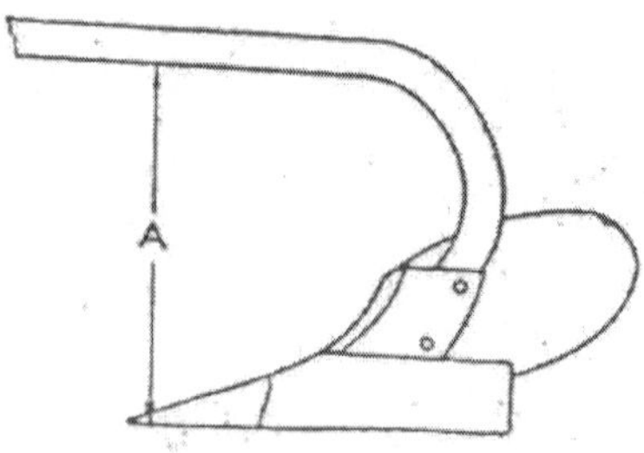

Fig. 13.12 Throat clearance of plough

Vertical clevis

It is a vertical plate with a number of holes at the end of the beam to control the depth of operation and to adjust the line of pull. The vertical clevis and line of pull is shown in Fig.13.11.

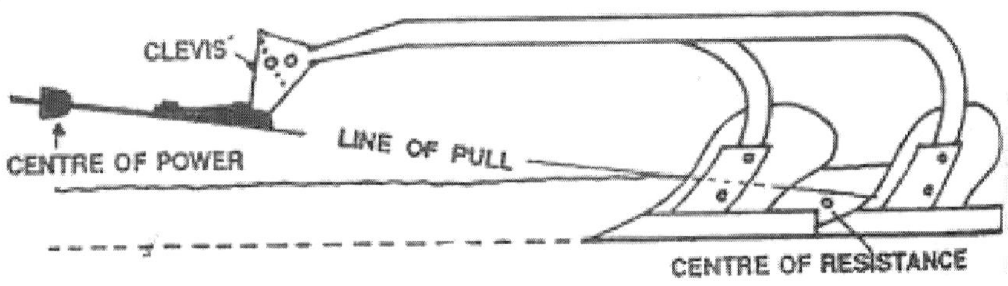

Fig. 13.13. Vertical Clevis and line of pull

HORIZONTAL CLEVIS

It is a device to make lateral adjustment of the plough relative to the line of pull.

Plough size

The perpendicular distance from wing of the share to the line jointing the point of the share and heel of land side is called size of plough. The size of the mould board plough is expressed by width of cut of the soil.

Tractor Drawn Implements

Tractor drawn implements possess higher working capacity and are operated at higher speeds. These implements need more technical knowledge for operations and maintenance work. Tractor drawn implements may be either trailed type or semi-mounted type or mounted type.

TRAILED TYPE IMPLEMENT

It is one that is pulled and guided from single hitch point but its weight is not supported by the tractor.

Semi-mounted type implement

This type of implement is one which is attached to the tractor only to the lower link of the hydraulic system. It is controlled directly by tractor steering unit but its weight is partly supported by the tractor.

Mounted type implement

A mounted implement is one which is attached to the three point linkage of hydraulic system, which can be controlled directly by the tractor steering unit. The implement is carried fully by the tractor. It can be raised, lowered and hold in any position.

TERMS USED IN PLOUGHING

Centre of power: It is the true point of hitch of a tractor.

Centre of resistance: It is the point at which the resultant of all the horizontal and vertical forces act. The centre lies at a distance equal to 3/4th size of the plough from the share wing.

Line of pull: It is an imaginary straight line passing from the centre of resistance through the clevis to the centre of pull.

Pull : It is the total force required to pull an implement.

(1) Fixed type (one way) mould board plough

One way plough throws the furrow slice to one side of the direction of motion and is commonly used everywhere. It may be long beam type or short beam type. There are single bottom, two bottom, three bottom, four bottom and five bottom fixed mould board ploughs are available which are based on the power source horse power.

Plate : 13.14 Two bottom fixed mould board plough

(2) Two-way or Reversible plough

It is a mould board plough which turns furrow slice to the right or left side of direction of travel as require. Such ploughs have two sets of opposed bottoms. In such plough, all the furrow can be turned towards the same side of the field by using one bottom for one direction of travel and the other bottom on the return trip. Two sets of bottom are so mounted that they can be raised or lowered independently or rotated along an axis. Two way ploughs have the advantages that they neither upset the slope of the land nor leave dead furrows or back furrows in the middle of the field.

Plate : 13.15 Single bottom reversible mould board plough

14

DISC PLOUGH

It is a plough which cuts, turns and in some cases breaks furrow slices by means of separately mounted large steel discs. A disc plough is designed with a view of reduce friction by making a rolling plough bottom instead of sliding plough bottom. A disc plough is shown in Plate 14.1. A disc plough works well in the conditions where mould board plough does not work satisfactorily.

Fig: 14.1 Disc plough

ADVANTAGES OF DISC PLOUGH

- A disc plough can be forced to penetrate into the soil which is too hard and dry for working with a mould board plough.
- It works well in sticky soil in which a mould board plough does not scour.
- It is more useful for deep ploughing.
- It can be used safely in stony and stumpy soil without much danger of breakage.
- A disc plough works well even after a considerable part of the disc is worn off in abrasive soil.

- It works in loose soil also (such as peat) without much clogging.

DISADVANTAGES OF DISC PLOUGH

- It is not suitable for covering surface trash and weeds as effectively as mould board plough does.
- Comparatively, the disc plough leaves the soil in rough and more clody condition than that of mould board plough.
- Disc plough is much heavier than mould board plough for equal capacities because penetration of this plough is affected largely by its weight rather than suction. There is one significant difference between mould board plough and disc plough i.e. mould board plough is forced into the ground by the suction of the plough, while the disc plough is forced into the ground by its own weight.

TYPES OF DISC PLOUGH

Disc ploughs are of two types' viz., Standard disc plough and Vertical disc plough.

(i) Standard disc plough

It consists of steel disc of 60 to 90 cm diameter, set at a certain angle to the direction of travel. Each disc revolves on a stub axle in a thrust bearing, carried at the lower end of a strong stand which is bolted to the plough beam Fig.14.1.

The angle of the disc to the vertical and to the furrow wall is adjustable. In action, the disc cuts the soil, breaks it and pushes it sideways. There is a little inversion of furrow slice as well as little burying of weeds and trashes. The disc plough may be mounted type or trailed type. In mounted disc plough, the side thrust is taken by the wheels of the tractor. Sometimes a rear wheel is fitted to take side thrust of the plough to some extent. In trailed type, side thrust is taken by the furrow wheel of the plough. Disc is made of heat treated steel of 5 mm to 10 mm thickness. The edge of the disc is well sharpened to cut the soil. The amount of concavity varies with the diameter of the disc. The approximate values being 8 cm for 60 cm diameter disc and 16 cm for 95 cm diameter. A few important terms connected with disc plough is explained below.

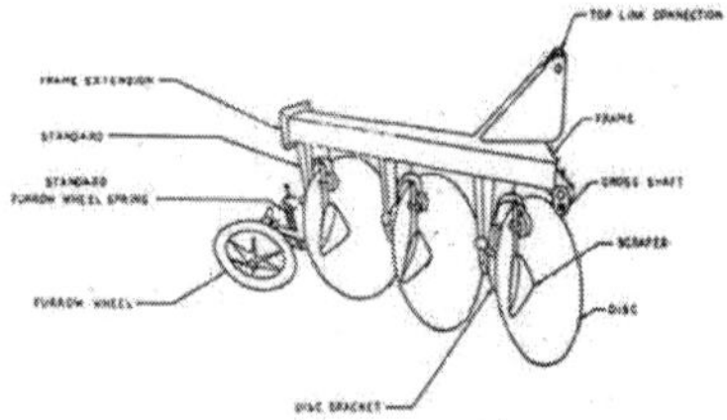

Fig. 14.2 Standard disc plough

Disc - It is a circular, concave revolving steel plate used for cutting and inverting the soil.

Disc angle - It is the angle at which the plane of the cutting edge of the disc is inclined to the direction of travel.. Usually the disc angle of good plough varies between 42° to 45°.

Tilt angle - It is the angle at which the plane of the cutting edge of the disc is inclined to a vertical line. The tilt angle varies from 15° to 25° for a good plough. The disc and tilt angles of disc plough are shown in Fig. 14.2.

Scraper – It is a device to remove soil that tends to stick to the working surface of a disc.

Concavity -It is the depth measured at the centre of the disc by placing its concave side on a flat surface.

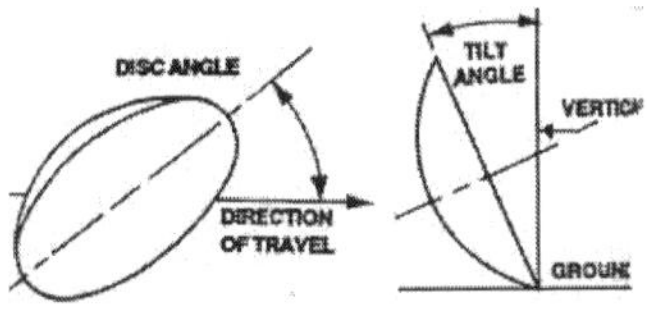

Fig. 14.3 Angles of disc plough

DRAFT OF DISC PLOUGH

The disc plough is lighter in draft than the mould board plough, turning same volume of soil in similar conditions. In very hard soil, some extra weight is added to the wheel which increases the draft. The bearings and scrapers of the disc plough also affect the draft.

Special purpose ploughs

1. Rotary tiller (Rotovator)

The rotary cultivator is widely considered to be the most important tool as it provides fine degree of pulverization enabling the necessary rapid and intimate mixing of soil besides reduction in traction demanded by the tractor driving wheels due to the ability of the soil working blades to provide some forward thrust to the cultivating outfit.

Fig: 14.4 Rotavator

The functional components include tynes, rotor, transmission system, universal joint, leveling board, shield, depth control arrangement, clutch and three point linkage connection. Rotary tiller is directly mounted to the tractor with the help of three point linkage. The power is transmitted from the tractor PTO (Power Take Off) shaft to a bevel gear box mounted on the top of the unit, through telescopic shaft and universal joint. From the bevel gear box the drive is further transmitted to a power shaft, chain and sprocket transmission system to the rotor. The tynes are fixed to the rotor and the rotor with tynes revolves in the same direction as the tractor wheels. The number of tynes varies from 28 -54. A leveling board is attached to the rear side of the unit for leveling the tilled soil. A depth control lever with depth wheel provided on either side of the unit ensures proper depth control.

The following types of blades are used with the rotovator.

i. 'L' type blade - Works well in trashy conditions, they are more effective in cutting weeds and they do not pulverize the soil much.

ii. Twisted blade - Suitable for deep tillage in relatively clean ground, but clogging and wrapping of trashes on the tynes and shafts.

iii. Straight blade - Employed on mulches designed mainly for secondary tillage.

The benefits of the rotary tiller are effective pulverization of soil ensures good plant growth, stubble and roots are completely cut and mixed with the soil and proper ground levelling after the operation.

2. Chisel plough

Chisel ploughs are used to break through and shatter compacted or otherwise impermeable soil layers. Deep tillage shatters compacted sub soil layers and aids in better infiltration and storage of rainwater in the crop root zone. The improved soil structure also results in better development of root system and the yield of crops and their drought tolerance is also improved. The functional component of the unit include reversible share, skids, beam, cross shaft and top link connection.

Chisel plough consists of heavy chisel type tyne which is pulled through the soil normally at a depth greater than that at which conventional ploughing would be done and bursting up the underlying layers of soil without bringing the sub-soil to the surface. The tynes of the implement are sturdy and strong enough to withstand the stresses applied when they are working at depth where the soil conditions are hardened. The implement frame is also strongly constructed usually of box section steel to withstand the stresses applied. The chisel plough has a sturdy but light structure made of 3 mm thick hollow rectangular tubular mild steel sections. The share has a lift angle of 20 degree, width of 25 mm and a length of 150 mm. The implement is protected by a shear pin, which prevents damage from over loading. The implement could be used for deep tillage upto a depth of

40 cm. The coverage is 0.42 ha/hr when operated at a spacing of 1.5 m between rows. The salient features of the unit are:

- The implement could be used for deep tillage upto a depth of 40 cm for bursting of the sub-soil hard pan, improving the drainage and aerating the soil.
- Reduces the bulk density of soil
- Two fold increase in hydraulic conductivity of sub-soil
- Conserves around 30 to 40% more soil moisture
- Roots proliferation is improved from 40 to 45%
- Easily operated by any 35 to 45 hp tractor

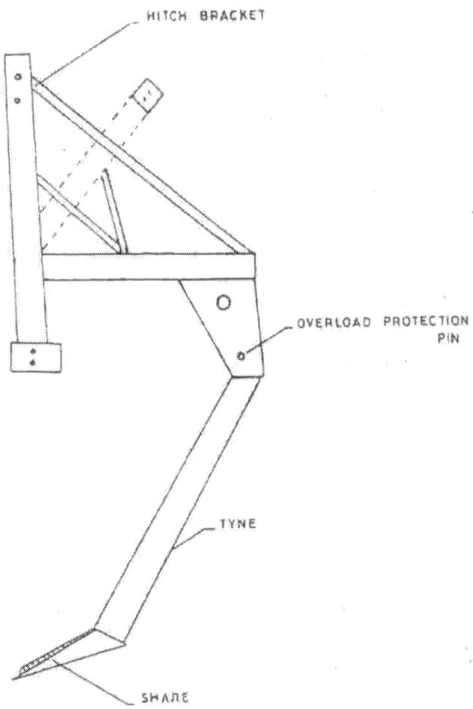

Fig. 14.5 Line diagram of chisel plough

3. *Sub-Soil plough*

The function of the sub-soil plough also called sub soiler is to penetrate deeper than the conventional cultivation machinery and break up the layers of the soil, which have become compacted due to the movement of heavy machinery or as a result of continuous ploughing at a constant depth. These compacted areas prevent the natural drainage of the soil and also inhibit the passage of air and nutrients through the soil structure. The sub- soiler consists of heavier tyne than the chisel plough to break through impervious layer shattering the sub-soil to a depth of 45 to 75 cm and requires 60 to 100 hp to operate it. The line diagram of sub soiler is shown in Fig. 14.4. The advantages are same as that of chisel plough.

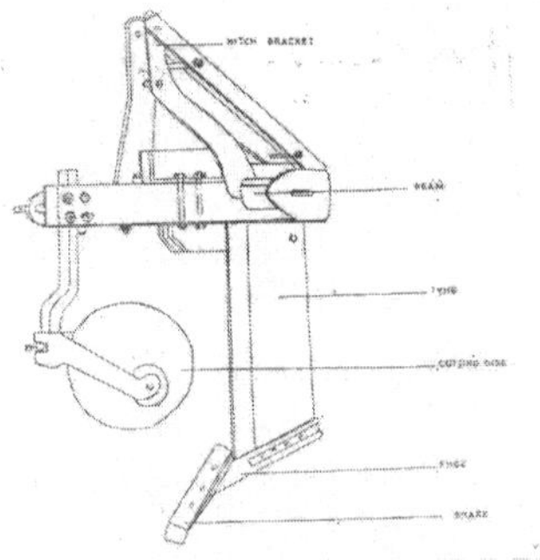

Fig. 14.6 Line diagram of Sub soiler

15

SECONDARY TILLAGE IMPLEMENTS

SECONDARY TILLAGE

Tillage operations following primary tillage which are performed to create proper soil tilth for seeding and planting are called secondary tillage.

These are lighter and finer operations, performed on the soil after primary tillage operations. Secondary tillage consists of conditioning the soil to meet the different tillage objectives. The implements used for secondary tillage operations are called secondary tillage implements. They include different types of harrow, cultivators, levelers, clod crushers and similar implements. These operations are generally done on the surface soil of the farm. Secondary tillage operations do not cause much soil inversion and shifting of soil from one place to other. These operations consume less power per unit area compared to primary tillage operations. The main objectives of secondary tillage operations are :

1. To break the big clods and to make the field surface uniform and leveled for sowing.
2. To cut crop residues and mix them with top soil of the field.

Secondary tillage implements may be tractor drawn or bullock drawn implements.

Bullock drawn implements include harrows, cultivators, hoes etc.

1. Harrow

Harrow is a secondary tillage implement that cuts the soil to a shallow depth for smoothening and pulverizing the soil as well as to cut the weeds and to mix the materials with the soil. There are several types of harrow used in India such as disc harrow, spike tooth harrow, spring tooth harrow, acme harrow, patela, triangular harrow, blade harrow,

2. Disc harrow

It is a harrow, which performs the harrowing operation by means of a set, or a number of sets of rotating slat discs, each set being mounted on a common shaft. Disc harrow is found very suitable for hard ground, full of stalks and grass fields. It cuts the lumps of soil, clods and roots. Discs are mounted on one, two or more axles which may be set at a variable angle to the line of motion. As the harrow is pulled ahead, the discs rotate on the ground. The disc harrow in field operation is shown in Plate 15.1.

Plate : 15.1 Disc harrow in operation

(Courtesy: http://www.woodsequipment.com)

Classification of disc harrow : The disc harrows are classified based on disc arrangement as Single action, Double action disc harrows, tandom disc harrows and Off-set disc harrows. Based on power source, they are classified as Tractor drawn and bullock drawn disc harrows.

a) Single action disc harrow

It is a harrow with two gangs placed end to end, which throw the soil in opposite directions. The discs are arranged in such a way that right side gang throws the soil towards right, and left side gang throws the soil towards left as shown in Fig. 15.1.

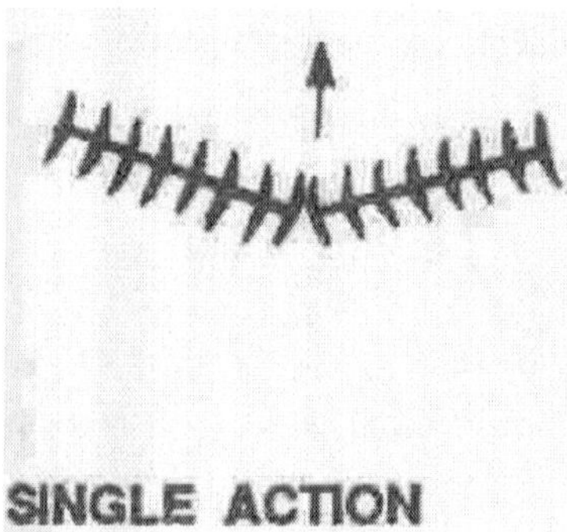

Fig. 15.2 line diagram of disc arrangement in single action disc harrow

b) Double action disc harrow

A disc harrow consisting of two or more gangs, in which a set of one or two gangs follow behind the set of the other one or two, arranged in such a way that the front and rear gangs throw the soil in opposite directions (Fig. 15.2). Thus the entire field is worked twice in each trip. It may be of two types i) Tandem and ii) Off-set type.

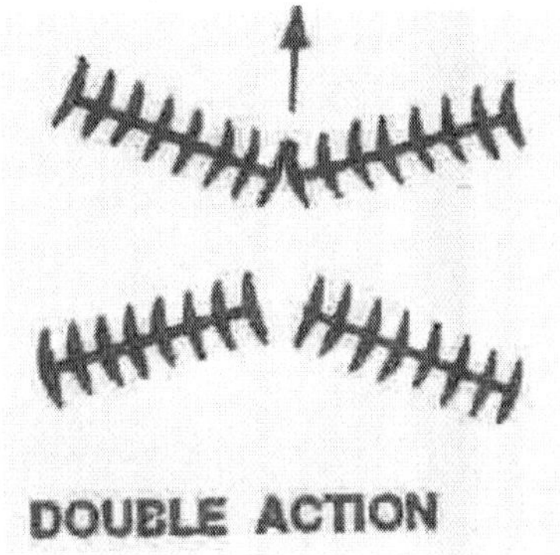

Fig. 15.3 line diagram of disc arrangement in double action disc harrow

c) Tandom disc harrow

It is a disc harrow comprising of four gangs in which each gang can be angled in opposite direction as shown in Fig.15.3.

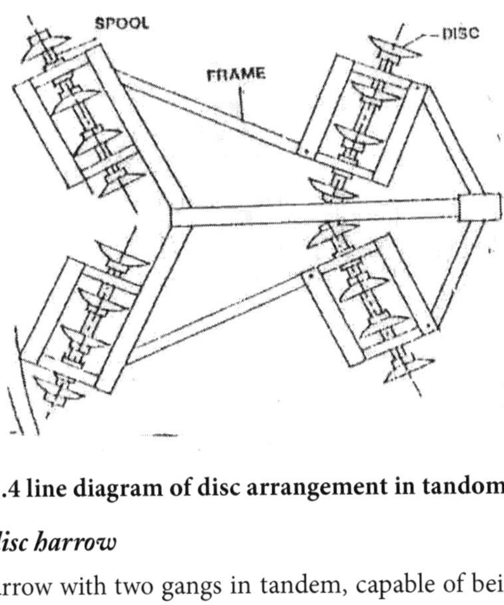

Fig. 15.4 line diagram of disc arrangement in tandom disc harrow

d) Off-set disc harrow

It is a disc harrow with two gangs in tandem, capable of being off-set to either side of the centre line of pull. Two gangs are fitted one behind the other. The soil is thrown in both directions because discs of both gangs face in opposite directions. It is very useful for orchards and gardens. It travels left or right of the tractor. The line of pull is not in the middle, that's why it is called off-set disc harrow. Off-set disc harrow is based on the basic principle that side thrust against the front gang is opposed by the side thrust of the rear gang. Hence the gangs are arranged at suitable angles so that both thrusts are counter balanced with each other.

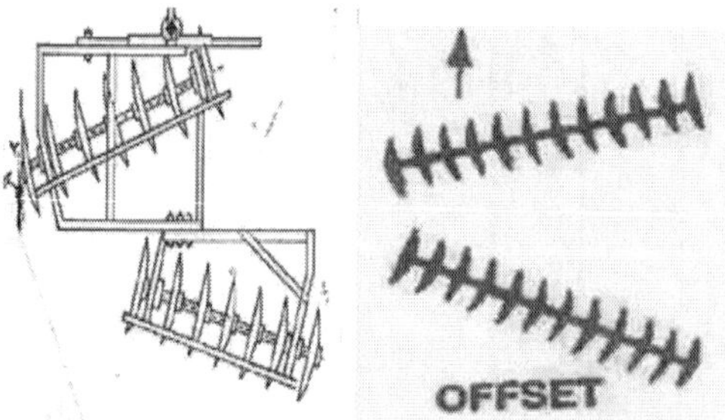

Fig. 15.5 line diagram of disc arrangement in offset disc harrow Components of a Disc harrow.

A disc harrow mainly consists of disc, gang, gang bolt, gang central lever, spool or spacer arbor bolt, bearings, transport wheels, scraper and weight box.

Disc: It is a circular concave revolving steel plate used for cutting a inverting the soil (Plate : 15.2). Disc is made of high glass heat-treated hardened steel. Tractor drawn disc harrows have concave discs of size varying from 35-70 cm diameter. Concavity of the disc affects penetration and pulverization of soil. Usually two types of disc are used in disc harrows, plain disc and cut away disc. Plain discs have plain edges and they are used for all normal works. Most of the harrows are fitted with plain discs only. Cut away discs have serrated edges and they cut stalks, grass and other vegetation. They are not effective for pulverization of soil but it is very useful for puddling the field especially for paddy cultivation.

Fig: 15.6 Different types of disc used in disc harrow

(Courtesy: http://www.sumoindia.com)

Gang: Each set of discs that are mounted on a common shaft is called the gang.

Gang bolt or arbor bolt: It is a long heavy of square headed bolt from the other end, a set of discs are mounted on the gang bolt. The spacing between the discs on the gang bolt ranges from 15 to 25 cm for light duty and 25 to 30 cm for heavy duty harrows. The angle between the axis of the gang bolt and the direction of travel is called the gang angle.

Gang control lever: A lever, which operates the gang mechanisms of the disc harrow, is called the gang control lever.

Spool or spacer: The flanked tube, mounted on the gang bolt between every two discs to prevent the lateral movement of the disc on the shaft is called the `spool' or `spacer'. It is just a device for keeping the discs at equal spacing on the gang bolt. It is usually cast in special shapes and sizes and is generally made of iron.

Bearing: Bearing is essential to counter act the end thrust of the gang due to soil thrust. The harrow bearings are subjected to heavy radial and thrust loads Chilled cast iron bearings are used for heavy radial and thrust loads and they are also used due to their durability.

Transport wheel: In trailing type disc harrow, the transport wheels are provided for transport work on roads and for preventing the edges of the discs from damage. Mounted type disc harrows do not require wheels for transport work.

Scraper: It prevents disc from clogging. It removes the soil that may stick to the concave side of the disc.

Weight box: A box like frame is provided on the main frame of the harrow for putting additional weight on the implement. Additional weight helps in increasing the penetration of the disc in the soil.

There are several factors which affect the penetration of disc harrow in the field. If the disc gangs are set perpendicular to the line of draft, the penetration is not adequate. Penetration can be increased by adding some additional weight on the frame of the harrow. For obtaining maximum penetration, the gangs should be set with the forward edges of the discs parallel to the direction of motion. If the hitch point is lowered, better penetration is achieved.

A sharp edged disc has more effective penetration compared to blunt edged disc. It is observed that penetration is better in low speed than in high speed. The following are a few adjustments for obtaining higher penetration

i. By increasing the disc angle

ii. By adding additional weight in harrow

iii. By lowering the hitch point

iv. By using the sharp edged discs of small diameter and lesser concavity

v. By regulating the speed.

SPIKE TOOTH HARROW

It is a harrow with peg shaped teeth of diamond cross section to a rectangular frame. It is used to break the clod, stir the soil, uproot the weeds, level the ground, break the soil and cover the seeds. Its principle is to smoothen and level the soil directly after ploughing. Spike tooth harrows may be of rigid type and flexible type.

The animal drawn spike tooth harrow is usually of rigid type. These may or may not have provisions for changing the angle of spikes in operating conditions. Tractor drawn harrows are usually flexible type (Fig.15.5). It has got the advantage of being turned up for transporting purpose. This harrow mainly consists of teeth, tooth bar frame, clamps, guard, braces, levers and hooks. The teeth are made up of hardened steel with square/ triangular/circular in section. The teeth are so placed on tooth bar that no tooth is directly behind the other. Teeth are fastened rigidly to the tooth bar. Clamps are rigidly fixed so as not to be loose while in operation.

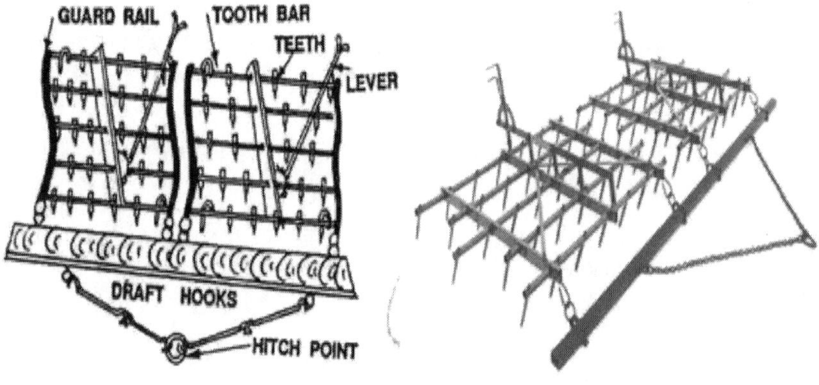

Fig. 15.7 line diagram of Spike tooth harrow Spring tooth harrow

It is a harrow with tough flexible teeth, suitable to work in hard and stony soils. Spring tooth harrow is fitted with springs having loops of elliptical shape as shown in Fig. 14.6. It gives a spring action in working condition. It is used in the soil when obstruction like stone, roots and weeds are hidden below the ground surface. This type pulverizes the soil and helps in killing weeds. This type of harrow mainly consists of teeth, tooth bar, clamps, frame, lever and links. Usually the teeth are made of spring steel. Sometimes reversible points are provided so that one end may be used after the other end is worn out. The teeth are fastened to the tooth bar by means of tooth clamps. They are provided to give rigidity and support to the harrow. The levers are provided for setting the teeth for varying the depth of harrowing. For light harrowing, the adjustment is done in slanting position. Draft hooks on each corner of every section for hitching purpose.

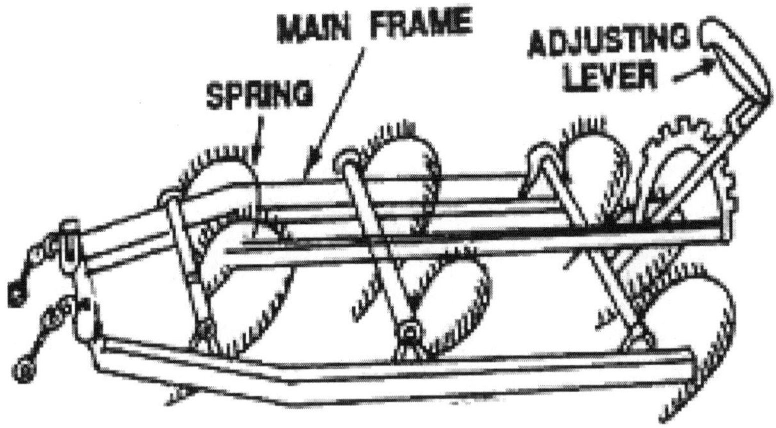

Fig. 15.8. Line diagram of spring tooth harrow

16

CULTIVATORS, LEVELERS, RIDGER AND BUND FORMER

CULTIVATOR

It is an implement for inter cultivation with laterally adjustable tines or discs to work between crop rows. This can be used for seed bed preparation and for sowing with seeding attachment. The times may have provision for vertical adjustments also.

The cultivator can be either Disc cultivator or Rotary cultivator or Tine cultivator.

Disc cultivator: It is a cultivator fitted with discs.

Rotary cultivator: It is a cultivator with tines or blades mounted on a power driven horizontal shaft as shown in Plate : 16.1.

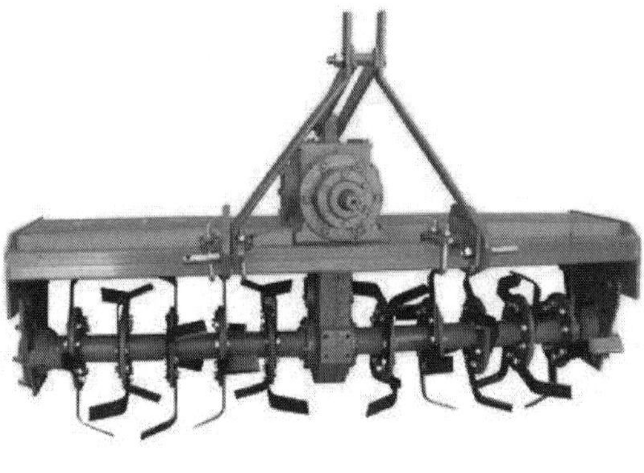

Fig : 16.1 Rotary cultivator

(Courtesy: http://www.diytrade.com)

Tine cultivator: It is a cultivator fitted with tines having shovels. The rigid tine cultivator is shown in Plate : 16.2.

Fig : 16.2 Rigid tyne cultivator

The cultivator stirs the soil, and breaks the clods. The tines fitted on the frame of the cultivator comb the soil deeply in the field. A cultivator performs functions intermediate between those of plough and the harrow. Destruction of weeds is the primary function of a cultivator.

The following are a few important functions performed by a cultivator.

- Interculture the fields.
- Destroy the weeds in the field.
- Aerate the soil for proper growth of crops.
- Conserve moisture by preparing mulch on the surface.
- To sow seeds when it is provided with sowing attachments.
- To prevent surface evaporation and encourage rapid infiltration of rain water into the soil.

Depending upon the type of power available for the implement, the cultivator can be classified as either Tractor drawn or Animal drawn. The tractor drawn cultivators are further classified as Trailed type or Mounted type.

Trailed type cultivator

It consists of a main frame which carries a number of cross members to which tines are fitted. At the forward end of the cultivator, there is a hitch arrangement for hitching purpose. A pair of wheels is provided in the cultivator. The life is operated by both wheels simultaneously so that draft remains even and uniform. The height of the hitch is adjusted so that main frame remains horizontal over a range of depth setting. The tines in each row are spaced widely to allow free passage of the soil and trash around them. The tines in subsequent rows are staggered so that the implement can cover the entire width nicely. The depth of working is set roughly by adjusting the tine in their clamps and the final depth control is done by a screw lever. Usually the tines are damaged due to turning the implement at the headland without lifting it up. Care should be taken to lift the tines off the ground before turning.

Mounted type Cultivator

Tractors fitted with hydraulic system operate the mounted type cultivators. A rectangular frame of angle iron is mounted on three point hydraulic linkage of the tractor. The cross members carry the tines in two staggered lines. For actual cutting the soil, different types of shovels and sweeps are used. A few important shovels and sweeps are

a) Single point shovel b) Double point shovel c) Spear head shovel d) Sweep e) Half sweep f)Furrower.

Depending upon the type of soil and crop, shovels are chosen for use on the cultivators. Usually tractor drawn cultivators are of two types, depending upon the flexibility and rigidity of tines (i) Cultivator with spring loaded tines (ii) Cultivator with rigid tines.

CULTIVATOR WITH SPRING LOADED TINES

A tine hinged to the frame and loaded with a spring so that it swings back when an obstacle is encountered, is called spring loaded tine as shown in Fig.16.1. Each of the tine of this type of cultivator is provided with two heavy coil springs pre tensioned to ensure minimum movement except when an obstacle is encountered. The springs operate, when the points strike roots or large stones by allowing the tines to ride over the obstruction, thus preventing damage. On passing over the obstruction, the tines are automatically reset and work continues without interruption. The tines are made of high carbon steel and are held in proper alignment on the main frame members. This type of cultivator is particularly recommended for soils which are embedded with stones or stumps. A pair of gauge wheel is provided on the cultivator for controlling the depth of operation. The cultivator may be fitted with 7, 9, 11, 13 tines or more depending upon the requirements.

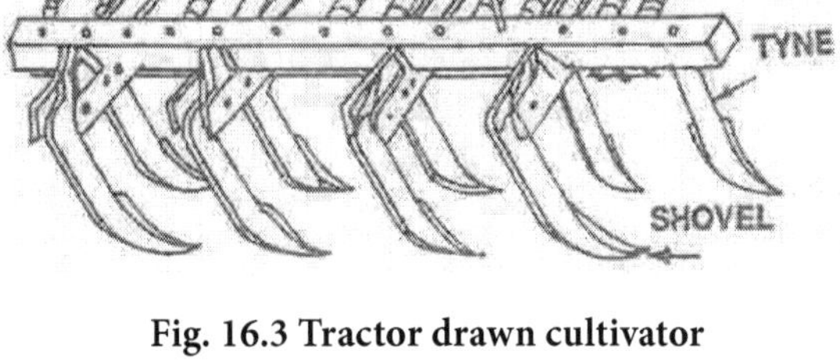

Fig. 16.3 Tractor drawn cultivator

CULTIVATOR WITH RIGID TINES

Rigid tines of the cultivators are those tines which do not deflect during the work in the field. The tines are bolted between angle braces, fastened to the main bars by sturdy clamps and bolts. Spacing of the tines are changed simply by slackening the bolts and sliding the braces to the desired position. Since rigid tines are mounted on the front and rear tool bars, the spacing between the tines can be easily adjusted without getting the tines choked with stubbles of the previous crop or weed growth. A pair of gauge wheel is used for controlling the depth of operation.

Duck foot cultivator

It is a type rigid cultivator which is used mostly for shallow ploughing, destruction of weeds and retention of moisture. It consists of steel frame and rigid tines to which sweeps are attached as shown in Plate : 16.2. The implement is attached to the tractor with three point hitch system and is controlled by hydraulic system. The sweeps are fabricated from high carbon steel. Number of sweeps can be adjusted according to requirement. Usually this cultivator is about 225 cm long; 60 cm wide with 7 sweeps.

Fig : 16.4 Duck foot cultivator

Terms used for evaluation of an implements Theoretical field capacity

It is the rate of field coverage of the implement, based on 100 per cent of time at the rated speed and covering 100 per cent of its rated width.

$$\text{Theoretical field capacity} = \frac{W \times S}{10} \text{ ha/hr}$$

Where,

W – Width of the implement in meters (m)

S - Speed of operation in Kilometers per hour (Km/hr)

Effective field capacity

It is the actual area covered by the implement based on its total time consumed and its actual width of operation.

Field efficiency

It is the ratio of effective field capacity and theoretical field capacity expressed in percent.

Effective field capacity Field efficiency =

Theoretical field capacity

Effective field capacity is calculated as follows S x W x E

C = 1000

Where,

C - Effective field capacity, hectare per hr (ha/hr). S - Speed of travel in km per hour.

W - Theoretical width of cut of the machine in metre, and E - Field efficiency in per cent.

Problem 1: A 5 x 20 cm double action disc harrow is operated by a tractor having a speed of 5 km/h. Calculate the actual field capacity, assuming the field efficiency of 80 percent.

Solution:

Size of the harrow (width) = 5 x 20 = 100 cm = 1 meter

S x W x E Actual field capacity, C = ----------------

1000

= (1 x 5 x 80) / 1000

Actual field capacity = 0.4 ha/hr.

Problem 2: A 3 x 30 cm plough is moving at a speed of 4 km/h. calculate how much time it take to plough 500 x 500 m field when the field efficiency is 70 %.

Solution:

Width of the plough = 3 x 30 = 90 cm = 0.9 m

S x W x E Effective field capacity, C = ----------------

1000

= (0.9 x 4 x 70)/ 1000

Effective field capacity, C = 0.25 ha/hr = 2500 m2/hr

(1 ha = 10,000 m2)

Time required = 500 x 500/ 2500 = 100 hr.

Problem 3: A 4 bottom 40 cm mould board plough is operating at 5.5 km/h speed with 75 % field efficiency . Calculate the rate of doing work in hectares per hour.

Solution :

Width of the plough = 4 x 40 = 160 cm = 1.6 m

Rate of doing work = 1.6 x 5.5 x 75/1000

Rate of doing work = 0.66 ha/h

Problem 4: An indigenous plough has a 20 cm wide furrow at the top and 10 cm depth. Calculate the volume of soil handled per day 8 hours if the speed of working is 2.5 km/h. Solution:

Furrow cross section = 10 x20/2

= 100 cm2

Distance traveled in 8 hours = 8 x 2.5 x 1000 = 20,000 m Volume of soil handled = 20000 x 100/ 10000 Volume of soil handled = 200 m3

****

LEVELERS

In irrigated areas land leveling is an essential operation of farming. Leveled fields receive uniform penetration of irrigation water with high efficiency. The possibility of water logging and soil erosion is reduced considerably. The entire leveled field becomes ready to receive timely agricultural operations like ploughing, seeding and interculturing without any delay. Smooth fields also facilitate the operation of field equipment and are highly desirable for using mechanical harvesters. Land leveling is usually done in the slack season when the field is free from crops, and the Labours and bullocks are idle.

Wooden logs or planks are the most common type of land levelers used by Indian farmers. They are operated in ploughed land to collect loose soil form high spots and to dump in into depressions. While operating in the field, the leveler is raised from the rear to take more cut and then it is tilted up side down to fill up the low spots elsewhere. The other improved type of land leveler which is used on the medium size farms, is called the leveling karaha or scraper. The scraper essentially consists of a bowl or 'U' shaped steel sheet made of 3 mm thick metal. Its cutting edge is generally made of high carbon steel. It is also provided with a wooden handle in the middle or two handles on the sides at the rear end. Provision is made at the front end for hitching. For taking the wear an extra steel sheet is welded underneath at the centre. The implement is pulled by a pair of bullocks. Two men are needed to operate it. One man controls the bullocks and the other man does the loading and unloading. The amount of work done in a day depends upon various factors such as hardness of the soil, transportation distance (lead) and volume of the soil cut each time. If the soil is hard, it is always helpful to loosen the soil first by ploughing or by any other means and then proceed with leveling. Tractor mounted type (plate : 16.4) or animal operated improved levelers are also used in India. The laser guided leveler operated by tractor is shown in Plate : 16.5. The laser guided leveler has a transmitter which is fitted outside the field to be leveled. The leveler is fixed with a receiver of the laser beam. It works automatically and the operator need not adjust the leveling blade.

Fig : 16.5 a Tractor drawn leveller

Fig : 16.5 b Laser guided leveler in operation

RIDGER

A ridger (Plate : 16.5) is an implement which cuts and turns the soil in two opposite directions simultaneously for forming ridges. It is also known as Furrower. Ridger is used to form ridges, for sowing row crop seeds and plants in well tilled soil. The ridger is also used for forming field furrows or channels, earthing up and similar other operations. Ridgers are also known ridging plough, middle buster plough and double mouldboard plough. The ridger generally has V-shaped or wedge-shaped share, fitted to the frog. The nose or the tip of the share penetrates into the soil and breaks the earth. The mouldboards lift, invert and cast aside the soil, forming deep channels and ridges of the required size.

(Courtesy: http://www.scotlandfarm.co.uk)

Plate : 16.6 Tractor operated ridger in operation

A Ridger required for row planted crops. It is also used for making field furrows or channels, earthing up, and similar other operations. It generally has a 'V' shaped share either rigidly fixed or hinged to the mould boards. The short beamed ridger is provided with a gauge wheel attached to the front end of the beam. This gauge wheel facilitates movement and control of the ridger. It may be operated by one or two pair of bullocks are a tractor. A ridger consists of beam, clevis, frog, handle, mouldboards, braces, share, and sliding shoe.

Bund former

It is used for making bunds or ridges by collecting the soil. Bunds are required to hold water in the soil, thereby conserve moisture and prevent run off. The size of the bund former is determined by measuring the maximum horizontal distance between the two rear ends of the forming boards. A bund former consists of forming board, beam and a handle.

Fig 16.7 Bullock drawn bund former

17

SOWING EQUIPMENT

Sowing or seeding is an art of placing seeds in the soil to have good germination in the field. A perfect seeding gives correct amount of seed per unit area, correct depth at which seed is placed in the soil and correct spacing between row-to-row and plant to plant.

METHODS OF SOWING

There are different methods of sowing such as Broadcasting, Dibbling, Drilling, Seed dropping behind the plough, Transplanting, Hill dropping and Check row planting.

(i) Broadcasting

Broadcasting is the process of random scattering of seed on the surface of seedbeds. It can be done manually or mechanically both. When broadcasting is done manually, uniformity of seed depends upon skill of the man. Soon after broadcasting the seeds are covered by planking or some other devices. Usually higher seed rate is obtained in this system. Mechanical broadcasters are used for large-scale work. This machine scatters the seeds on the surface of the seedbed at controlled rates.

(ii) Dibbling

Dibbling is the process of placing seeds in holes made in seedbed and covering them. In this method, seeds are placed in holes made at definite depth at fixed spacing. The equipment used for dibbling is called dibbler. It is a conical instrument used to make proper holes in the field. Small hand dibblers are made with several conical projections made in a frame. This is very time consuming process, so it is not suitable for small seeds. Mostly vegetables are sown in this way.

(iii) Drilling

Drilling consists of dropping the seeds in furrow lines in a continuous flow and covering them with soil. Seed metering may be done either manually or mechanically.

The number of rows planted may be one or more. This method is very helpful in achieving proper depth, proper spacing and proper amount of seed to be sown in the field. Drilling can be done by (1) Sowing behind the plough (2) Bullock drawn seed drills (3) Tractor drawn seed drills.

(iv) Seed dropping behind the plough

It is very common method used in villages. It is used for seed like maize, gram, peas, wheat and barley. A man drops seeds in the furrow behind the plough. Sowing behind the plough can be done by a device known as malobansa. It consists of a bamboo tube provided with a funnel shaped mouth. One man drops the seeds through the funnel and other man handles the plough and the bullocks. This is a slow and laborious method.

(v) Transplanting

Transplanting consists of preparing seedlings in nursery and then planting these seedlings in the prepared field. It is commonly done for vegetable and flowers. It is very time consuming operation. Equipment for placing plants in the soil is called transplanter.

(vi) Hill dropping

In this method, seeds are dropped at fixed spacing and not in a continuous stream. Thus the spacing between plant to plant in a row is constant. In case of drills, the seeds are dropped in continuous stream and the spacing between plant to plant in a row is not constant.

(vii) Check row planting

It is a method of planting, in which row-to-row and plant-to-plant distance is uniform. In this method, seeds are planted precisely along straight parallel furrows. The rows are always in two perpendicular directions. A machine used for check row planting is called check row planter.

SEED DRILL

Seed drill is a machine for placing the seeds in a continuous flow in furrows at uniform rate and at controlled depth with or without the arrangement of covering them with soil.

The different functions of seed drill are:

(i) To carry the seeds.

(ii) To open furrow to an uniform depth

(iii) To meter the seeds

(iv) To place the seed in furrows in an acceptable pattern

(v) To cover the seeds and compact the soil around the seed.

Seed cum fertilizer drill

Seed drills, fitted with fertilizer dropping attachment, distribute the fertilizer uniformly on the ground. It is called seed cum fertilizer drill (Plate : 17.1 and 17.2) Such a drill has a large seed box which is divided lengthwise into two compartments, one for seed and another for fertilizers. Seed drill may be classified as (i) Bullock drawn (ii) Tractor drawn.

Fig: 17.1 a Tractor drawn Seed cum fertilizer drill

Fig: 17.1 b Seed cum fertilizer drill in operation

Depending upon the method of metering the seeds, bullock drawn seed drill can be further divided into two groups based on the type of seed dropping as by hand or mechanically. There are a number of bullock drawn implements which are used for sowing seeds in which seeds are dropped by hand. The most popular implement is three lined cultivators with seeding attachment. In different parts of the country it is made in different sizes and shapes.

Components of seed drill

A seed drill with has mechanical seed metering device mainly consists of :

(i) Frame (ii) Seed box (iii) Seed metering mechanism (iv) Furrow openers (iv) Covering device and (vi) Transport wheels.

FRAME

The frame is usually made of angle iron with suitable braces and brackets. The frame is strong enough to withstand all types of loads in working condition.

Seed box

It may be made of mild steel sheet or galvanized iron with a suitable cover. A small agitator is sometimes provided to prevent clogging of seeds.

Covering device

It is a device to refill a furrow after the seed has been placed in it. Covering the seeds are usually done by patta, chains, drags, packers, rollers or press wheels, designed in various sizes and shapes.

Transport wheel

There are two wheels fitted on the main axle. Some seed drills have got pneumatic wheels also. The wheels have suitable attachments to transmit power to operate seed dropping mechanism.

Seed metering mechanism

The mechanism of a seed drill or fertilizer distributor which deliver seeds or fertilizers from the hopper at selected rates is called seed metering mechanism. There are a number of seed metering mechanism such as : Fluted roller feed type, internal double run type, cup feed type, cell feed mechanism, brush feed mechanism, auger feed mechanism, picker wheel mechanism and star wheel mechanism.

(a) Fluted roller feed type

It is a seed metering device with adjustable fluted roller to collect and deliver the seeds into the seed tube. Fluted roller feed type mechanism (Fig.17.1) consists of a fluted wheel, feed roller, feed cut-off and adjustable gate for different sizes of grains.

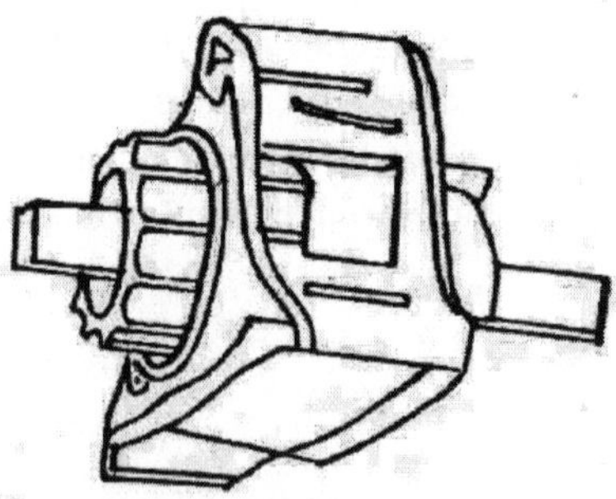

Fig. 17.2 Fluted roller type

The feed roller and the feed cut-off device are mounted on a shaft, running through the feed cups. The roller carries grooves throughout its periphery. It rotates with the axle over which it is mounted throws the grains out on the adjustable gate from where it falls into the seed tube. The fluted rollers which are mounted at the bottom of the seed box, receive seeds into longitudinal grooves and pass on to the seed tube through the holes provided for this purpose. By shifting the fluted wheel side ways, the length of the grooves exposed to the seed can be increased or decreased and hence the amount of seed is controlled. Thus a number of selection is available between closed position and full exposure of fluted wheel. The fluted feed mechanism is more positive in its metering action than the internal double run method.

(b) Internal double run type

It is a seed metering device in which the feed wheel is provided with fine and coarse ribbed flanges as shown in Fig. 17.2. It consists of discs, mounted on a spindle and housed in a casing fitted below the seed box. It has double faced wheel. Internal double- run type roller one face has a larger opening for larger seeds and the other face has smaller opening for smaller seeds. A gate is provided in the bottom-of the box to cover the opening not in use. The rate of seeding is varied by adjusting the speed of the spindle which carries the discs.

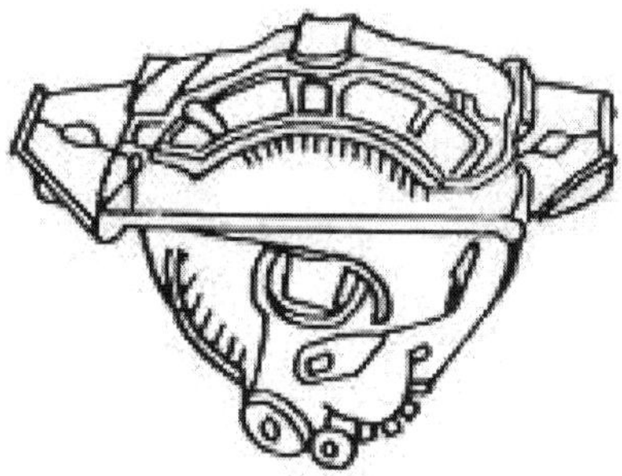

Fig. 17.3 Internal double run type

(c) Cup feed mechanism

It is a mechanism consisting of cups or spoons on the periphery of a vertical rotating disc which picks up the seeds from the hopper and delivers them into the seed tubes as shown in Fig.17.3. It consists of a seed hopper which has two parts. The upper one is called grain box and the lower one is called feed box. The seed delivery mechanism consists of a spindle, carrying a number of discs with a ring of cups attached to the periphery of each disc. The spindle with its frame and attachment is called seed barrel. When the spindle rotates, one disc with its set of cups rotates and picks up few seeds and drops them into small hoppers. The cups have two faces, one for larger seeds and the rate at which the seed barrel revolves. This type of mechanism is common on British seed drills.

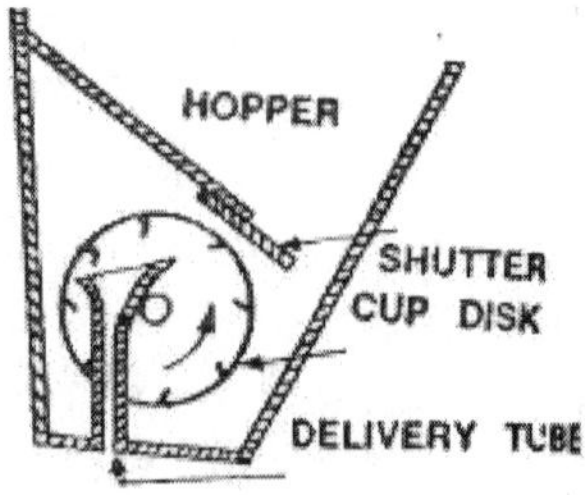

Fig. 17.4 Cup feed mechanism

(d) Cell feed mechanism - It is a mechanism in which seeds are collected and delivered by a series of equally spaced cells on the periphery of a circular plate or wheel.

(e) Brush feed mechanism - It is a mechanism in which a rotating brush regulates the flow of seed from the hopper. A number of bullock drawn planters in the country have Brush feed mechanism.

(f) Auger feed mechanism - It is a distributing mechanism, consisting of an auger which causes a substance to flow evenly in the field, through an aperture at the base or on the side of the hopper. Many of the fertilizer drills on the country have got Auger feed mechanism.

(g) Picker wheel mechanism - It is a mechanism in which a vertical plate is provided with radially projected arms, which drop the large seeds like potato in furrows with the help of suitable jaws.

(h) Star wheel mechanism - It is a feed mechanism which consists of a toothed wheel, rotating in a horizontal plane and conveying the fertilizer through a feed gate below the Star wheel.

FURROW OPENERS

The furrow openers are provided in a seed drill for opening a furrow. The seed tube conducts the seed from the feed mechanism into the boot from where they fall into the furrows.

Type of furrow openers

There are different type of furrow openers, which are: a) Shovel type, b) Shoe type and c) Disc type (single disc, double disc).

(a) Shovel Type

Shovel type furrow opener are the widely used in seed drills. There are three of shovels in use. They are: (a) reversible shovel (b) single point shovel and (c) spear point shovel.

Shovel type openers are best suited for stony or root infested fields. These shovels are bolted to the flat iron shanks at the point where boots are fitted which carry the end of the seed tubes. In order to prevent shock loads due to obstructions, springs are provided. It is easy in construction, cheaper and easily repairable. It is very common with usual seed drill.

(b) Shoe Type

It works well in trashy soils where the seed beds are not smoothly prepared. They are made from two flat pieces of steel welded together to form a cutting edge. It is specially suited for black cotton soil. Shoe is made of carbon steel having minimum carbon content of 0.5 per cent with a minimum thickness of 4 mm.

(c) Disc Type

They are two types: (i) Single disc type and (ii) Double disc type.

(i) Single disc type

It is furrow opener consisting of one concave disc. Disc type furrow openers are found suitable where plant debris or trash mulches are used. It consists of a curved disc made of hardened steel. It is set at an angle which while operating, shift the soil to one side making a small ridge. The disc is kept clean by two scrapers, one toe shaped at the convex side and one 'T' shaped at the concave side. It works in sticky soils also, but the discs are costly and maintenance work is a bit difficult.

ii) Double disc type

In double disc type furrow opener there are two flat discs, set at an angle to each other. The disc opens a clean furrow and leaves a small ridge in the center. The seeds are dropped between the two discs, providing more accurate placement. It is suitable for trashy lands. Seed drills attached with tractors having high speeds, usually use this type of furrow opener.

The furrow opener consists of Tine, shovel, seed tube and boot for seed and fertilizer.

(1) Tine: A thin pointed projection of a implement such as fork or pitchfork

(2) Shovel- It is made of carbon steel having carbon content of 0.5 per cent and a minimum thickness of 4.0 mm.

(3) Seed tube - It is a tube which carries the seeds from the metering device to the boot. Seed tubes are provided at the lowest lines through suitable boots and furrow openers. The most common to the furrow lines through suitable boots and furrow openers. The most common type of seed tube is the steel ribbon one. The ribbon is rolled in the form in such a way that lower edge of the tube is thinner than the edge. Polythene or rubber tubes are also used of this purpose. The minimum diameter of seed and fertilizer tube is 25 mm.

(4) Boot - It is a part of the sowing machine which conveys the seeds or fertilizers from the delivery tube to the furrow. It is bolted or welded to the tine. It is a harrow

casting, into which the lower end of the seed tube is inserted and to which the furrow openers are attached.

PLANTER

Planter is normally used for those seeds which are larger in size and can not be sown by usual seed drills. The main functions of the planter are:

(i) To open the furrow

(ii) To meter the seed

(iii) To deposit the seed in the furrow

(iv) To cover the seeds and compact the soil over it.

A planter consists of hopper, seed metering device, knock out arrangement, cut- off mechanism, furrow opener and other accessories.

Hopper : A planter has seed hopper for each row. Hopper is usually made of mild steel or any other suitable material.

Seed metering device in a planter : There are a number of seed metering devices in a planter but the most common device consist of a rotating plate at the bottom of seed hoppers. In some planters, vertical rotors as well as inclined rotors are also used. The most common is the horizontal seed plate used in planters. The horizontal seed plates have got suitable notches or holes called cell. Depending upon the type of notches on the plates, it is of three types. (i) Edge drop (ii) Flat drop (iii) Hill drop. The edge drop carries the seed on edge in the cell of the plate. The flat drop carries the seed on a flat in the cell of the plate. Only one seed is allowed in the cell at each time. In hill drop, the cells round the edge of the plate are large enough to admit several seeds at a time.

The rotating plate receives the seeds from the hopper. The plate moves under an arrangement called cut-off which allows only those seeds which are accommodated in the cells. Cut-off mechanism cuts-off or brushes out excess seeds from the cells of the feed mechanism.

Knock out arrangement : This mechanism is a device which knocks out the seeds from the cells or picker heads of the mechanism. It consists of rollers, star wheels or rounded points which are forced into the cells by the pressure of a spring and eject seeds out of the cells.

The spacing of seeds or hills in the row is determined by the ratio of linear or peripheral speed of the cells to the forward speed of the planter and by the distance between the cells in the metering unit. The accuracy of the planter depends upon several factors such as: (i) speed of seed plate (ii) shape and size of cells (iii) shape of hopper bottom and (iv) uniformity of seed size.

Planter is usually used for those seeds which are required to be sown at equal intervals between plant to plant.

Calibration of seed drill : The procedure of testing the seed drill for correct seed rate is called calibration of seed drill. It is necessary to calibrate the seed drill before putting it in actual use to find the desired seed rate. It is done to get the pre determined seed rate of the machine. The following steps are followed for calibration of seed drill.

Procedure :

i. Determine the nominal width (W) of seed drill .

W = M x S,

Where, M = Number of furrow openers

S = Spacing between the openers, m

ii. Find the length of the strip (L) having nominal width (W) necessary to cover 1/25 ha (1 /25 x 10000 m2).

L = 400/W, meter

iii. Determine the number of revolutions (N) of the ground wheel of the seed drill required to cover the length of the strip (L).

L = x D x N = 400/W

N =400/x D x W revolutions per minute

iv. Jack the seed drill so that the ground wheels turn freely. Make a mark on the drive wheel and a corresponding mark at a convenient place on the body of the drill to help in counting the revolutions of the ground wheel..

v. Fill the selected seed in the seed hopper. Place a container under each boot for collecting the seeds dropped from the hopper.

vi. Set the seed rate control adjustment for maximum position and mark this position on the control for reference.

vii. Engage the clutch and rotate the ground wheel for N = 400/P x D x N, revolutions per minute.

viii. Weigh the quantity of seed collected in the container and record the observation.

ix. Calculate the seed rate in kg/ha.

x. If the calculated seed rate is higher or lower than the desired rate of selected crop, repeat the process by adjusting the seed rate control adjustment till the desired seed rate is obtained.

Problem 1 : Calculate the cost of seeding one hectare of land with bullock drawn seed drill of 5 x 22 cm size. The speed of bullocks is 3 km/hr. Hire charge of bullocks Rs.200/- per pair, hire charges of seed drill Rs.50/- per day and wage of operator Rs.150/- per day of 8 hours.

Solution :

Width of seed drill = 5x22 = 110 cm = 1.1 m

Area covered/hr = width x speed = 1.1 x 3x1000 m = 0.33 ha Time taken/ha = 1/0.33 = 3.03 hrs.

Hiring cost of per hour = 200+50+150 = Rs 50.8

Cost of seeding per hectare = 3.03 x 50= Rs. 151.50

Potato planter

Planting equipment for commercial crop like Potatoes were considered essential due to excessive labour requirements in traditional methods of planting these tubers and, therefore, few machines have been developed and commercialised.

Potato tubers are planted either on flat bed or on ridgers or in small furrows depending upon the soil types and moisture content of soils. While planting potato tubers on flats or in furrows, the ridging operation is carried out at later dates. In improved method of potato planting, fertilizers are placed at the time of planting at a depth greater than that of the tubers. An improved potato planter cum fertilizer applicator is meant for drilling the fertilizer, dropping of tubers on flat surface and subsequently covering those with soils making a small ridge in one operation. However, the machine has not become a commercial unit for a wide spread acceptance.

Tractor operated potato planter-cum-cultivator is a commercial machine. It is used for planting tubers with 60 cm spacing of rows. However, the plant to plant spacing can be adjusted between 20 and 45 cm distance. The graded tubers of 2.5 to 4 cm size weighing 50-60 gm each are suitable for machine planting. By removing the hopper, the furrow opener and power transmitting mechanism, the machines suits well for earthing operations. Ridgers are so adjusted that earthing operation suits the row spacing.

18

CROP PROPAGATION TOOLS, CUTTING, GRAFTING, PRUNING TOOLS PLANT PROPAGATION

Most plants reproduce more of their kind through production of seed. This is Sexual reproduction and it involves the exchange of genetic material between tow parent plants. Many ornamental plants do come "true" from seed.

To increase the numbers of these plants, gardeners and horticulturists use Asexual propagation. In asexual propagation, the new plants are genetically exact copies or clones of a single parent plant. The methods used in asexual propagation range from taking leaf cutting of African violets to grafting apple cuttings onto root stocks.

Techniques for vegetative propagation : There are different techniques for vegetative propagation such as Air or ground layering, Division, Grafting and bud grafting which is widely used in fruit tree propagation, Micro propagation, Stolen or runners, Storage organs such as bulbs, corms tubers and rhizomes, Striking or cuttings and Twin-scaling.

Fruit tree propagation is usually carried out through asexual reproduction by grafting or budding the desired variety onto a suitable rootstock.

Grafting is a method of asexual plant propagation widely used in agriculture and horticulture where the tissues of one plant are encouraged to fuse with those of another. It is most commonly used for the propagation of trees and shrubs grown commercially. (Grafting is limited to dicots and gymnosperms. Monocots lack the vascular cambium required).

In most cases, one plant is selected for its roots, and this is called the stock or rootstock. The other plant is selected for its stems, leaves, flowers, or fruits and is called the scion. The scion contains the desired genes to duplicated in future production by the stock/scion plant.

In stem grafting, a common grafting method, a shoot of a selected, desired plant cultivar is grafting onto the stock of another type.

In another common from called budding, a dormant side bud is grafting on the stem of another stock plant, and when it has fused successfully, it is encouraged to grow by cutting out the stem above the new bud.

For successful grafting to take place, the vascular cambium tissues of the stock and scion plants must be placed in contact with each other. Both tissues must be kept alive until the graft has taken, usually a period of a few weeks. Successful grafting only required that a vascular connection takes place between the two tissues. A physical weak point after still occurs at the graft, because the structural tissue of the two distinct plants, such as wood may not fuse.

REASONS FOR GRAFTING AND BUDDING

1. Change varieties or cultivars
2. Optimize cross-pollination and pollination.
3. Take advantage of particular rootstocks
4. Benefit from inter stocks
5. Perpetuate clones
6. Produce certain plant forms Repair damaged plants
7. Increase the growth rate of seedlings
8. Index viruses

Grafting tools

The grafting tool is a simple product, which will allow to make flawless grafts. The grafting tool can graft all kinds of plants, and is particularly well suited to grafting fruit trees. The clean cuts made by the tool allow to get a perfect graft, every time.

The main features of the grafting tool are:

- Easy to use - Create precise grafts quickly.
- Trimming blades - prepare stock for grafting.
- Replaceable grafting blade - make sure you always get a clean cut.
- Use of high carbon steel blades - longer lasting strength.
- Quickly and easily create professional quality grafts.
- The clean cuts made by the grafting tool should allow to get a perfect fit, every time.
- This increases the likelihood of the graft 'taking', leading to much less wastage.

- The 2 in 1 tool contains both a pruner and grafting guillotine. This allows to first prepare the plants to be grafted, by trimming away excess leaves and twigs. Then using the special grafting guillotine one can make a perfect cut into the grafting stock.
- Ergonomic handles: The handle should be ergonomically designed for the operator.

Tools used for budding and grafting : The different simple tools used are Dibber, Budding and grafting knives, Grafting Tools, Grafting Tape, Pruning and lopping shears

Dibber : Dibber are used to make holes in seed bed to plant seeds, seedlings and bulbs. They are ergonomically designed consisting of a stainless steel sharp dibber and a hard wood handle joined together using a ferrule. The sharp dibber glides effortlessly into all soils.

Grafting and Budding knives : Budding and grafting knives are designed specifically for these purposes and should not be used for carving and whittling wood. Using them for other purposes will make cutting edge blunt very quickly. Grafting and budding knives must be kept razor sharp so they will cut smoothly. They are available in either left or right-handed models. The blade is beveled on only one side, unlike conventional knives, which have blades that bevel on both sides down to the cutting edge.

Grafting Tools - Grafting tool is designed for making the cleft graft. It is used when the rootstock's diameter is greater than 1 inch. The wedge-shaped blade is used to split the stock, and the flat pick opens the cleft so that scions can be inserted. Once in place, the flat pick is removed and the cleft comes together to hold the scions in position.

Grafting Tape - It can be used for many things… Grafting…utilized as a cover/ protecting bandage when using a splint for broken/cracked branches, twigs. It is also used to hold the soil and roots for the root over rock design of Bonsai styling.

Pruning and Lopping Shears : Pruning and Lopping Shears are the scissors or sliding blade type rather than the blade and anvil type. If use to harvest scion wood or bud sticks, blade and anvil pruner will crush plant tissue. As with knives, pruning and lopping shears should be kept razor sharp to give clean, close cuts.

PRUNING TOOLS

Plants differ from other living things in one important respect: one can improve a plant's vigor by removing healthy parts of it. This is known as pruning. It is a process of carefully removing the parts of the plant - even healthy parts - to make it healthier still or more attractive, or to improve other qualities, such as its fruit-bearing ability. It's important to point out that the emphasis here is on improvement, not removal. While tools like chain saws can be effective pruning devices, they are principally designed for large-scale removal.

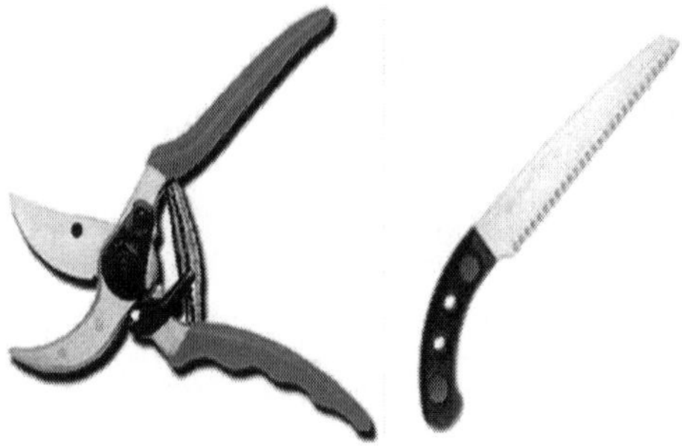

Fig : 18.1 Secateur & Saw tooth knife

(Courtesy: http://4.bp.blogspot.com & http://www.manualpolesawsviews.tk)

There are two types of pruning tools such as Anvil and Bypass. Anvil pruners and loppers have a blade that closes against an anvil on the lower jaw. The anvil is a softer metal than the blade. Bypass pruners and loppers have a blade that sweeps past the lower jaw. There's an important difference between the two. Anvil loppers and pruners tend to crush the material that is being cut and are used where cleanliness of cut is not as important as removal. This occurs when cutting dead material or making preparatory cuts - those cuts that occur before the final cut is made. For example, in some cases to shorten a branch before removing it. The preparatory cuts are those that reduce the branch's length before making the final cut. At other times, naturally fallen or pruned branches will need to be cut into small lengths for disposal. The anvil pruners and loppers are used to make all these cuts, as well as for thinning brush.

A bypass pruner or lopper is reserved for cuts that affect the health of the plant. It should be kept razor sharp - again, save the grunt work for anvil pruners and loppers. Finally, pruning tools that are used in one hand are known as pruners (or hand shears) while large tools that require you to use both hands are known as loppers.

19

INTER-CULTURE EQUIPMENT

The intercultural equipments are mainly used for removing the weed and control of weeds. Weeds grow along with the plants and compete with them for their nutrients. The weeds growing along with crops compete for moisture, light and nutrients. Hence, it is essential to remove them.

The following weeding tools and implements are used by the farmers.

1. Traditional hand tools.
2. Traditional and Improved hoes.
3. Wheel hoes.
4. Animal drawn multipurpose hoe.
5. Tractor drawn interculture equipment.
6. Power weeders.

Among the traditional hand tools, the Khurpi and the weeding hooks are popular tools made by local antisense for the use on small and marginal farms. Khurpi is used in India may vary in their size: shape and weight, but they have common basic parts i.e., a cutting blade and a small wooden handle for the grip. A person used it while squatting on the ground. The khurpi with a long narrow blade is preferred for weeding around the flower plants, broad casted crops and vegetable crops. One with wider blades is generally used for earthing up operations or for weeding crops sown in lines. However, a man covers about 0.025 hectare in a day under normal working condition.

The hoe is a versatile form of implement used for many operations, i.e. seed bed preparation, ridge making, channel shaping and weeding. It is also used for removing plant roots, harvesting root crops and thinning drilled crops. The two common types of hoes used by Indian farmers are hand hoes and animal drawn hoes.

Hand hoes are used to cultivate small area of land by human labour. Among the indigenous types of hand hoes, the Kodali (narrow spade) is the most popular one. Kodali is similar to a phawara (broad spade), the difference being that instead of a wide thin cutting blade, a narrow long pointed thicker section blade is attached to the handle. The person working with it has to bend his body. It is used for inter cultivating maize and sugarcane crops, and for earthing up the potato crops sown in lines. About 0.04 hectare can be covered in a day by one man. An improved hand hoe is operated in the standing position. It is provided with a long handle fitted in the middle of the cutting blade. One end of the blade is about 10 cm wide sharp edge and the other end is pointed narrow one for making small furrows. It can be used for cultivating and weeding very close to the individual plant.

The Grubber is a manual pull type hoe suitable for weeding and inter culture of upland row crops in black cotton soil regions. It is provided with 3 blades and the field capacity is 0.005 ha per hour.

The wheel hoe is another implement which is used for cultivating the land between rows. It consists of a wheel, two handles and a tyne to place the cutting tool on. Either a reversible shovel or a 3 prong fork or rake or a sweep is used as cutting tool, depending upon the weeds and moisture condition. A man operates the hoe in standing position by pushing through a short length each time. In a working day, about 0.04 hectare can be covered. Animal drawn weeding implements are pulled either by single animal or a pair of animals. These implements may either be single row type or multi row units. The three tyne cultivator or 'Triphali', Akola hoe, Bardole hoe or two 'Blade hoe' are the most popular implements in different regions for row crop interculture operations. It is essential to provide wider spacing (above 30 cm) for movement of animals and implement if animal drawn weeders are to be used.

The main parts of the blade hoe are:

1. Head piece
2. Prong
3. Blade
4. Handle and
5. Beam

The prongs make an angle of about 45o downward with the horizontal plane. At the end of each prong, the blade is attached. It loosens the upper surface of the soil and is generally used for interculturing sorghum (jowar), cotton, groundnut and other kharif crops. The hoe width is maintained between 25 and 75 cm depending upon the size of the bullocks and type of soil.

DRY LAND WEEDER

Dry land weeders with long handles are suitable for weeding in row crops in rainfed and garden lands. The long handle eliminates the back strain and provides comfort to the operator for continuous operation in standing posture. This is a long handled tool and consists of 25 mm diameter, 1200 mm long conduit pipe over which 520 mm long handle is fitted (Fig. 19.1). To the bottom of the vertical pipe frame, two arms made of 250 x 25 x 3 mm of MS plates are fitted. At the extreme end of the arm 120 mm diameter star wheel is fixed. The cutting blade is fitted to the bottom portion of the arm, and 200 mm to the back of the star wheel. The star wheel facilitates easy movement of the tool. The cutting blade cuts the weeds. The operating width of the blade is 120 mm. The star type is suitable for operation in loamy and sandy soil. In the peg type the star wheel is replaced by pegs facilitating easier operation in clay soil. The area covered by the weeder is about 0.08 ha/ day.

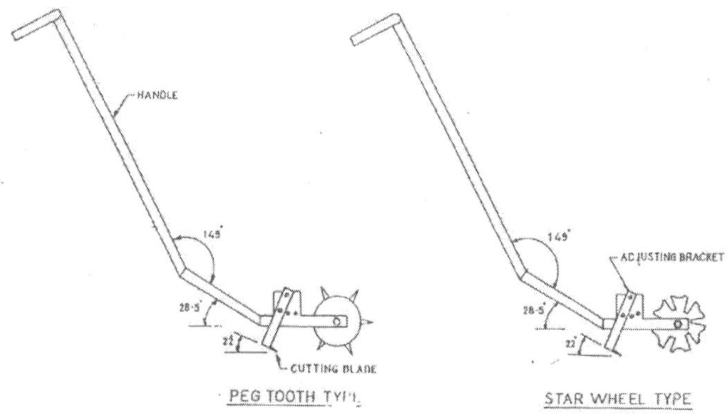

Fig. 19.1 Dry land weeders

ENGINE OPERATED WEEDER

Usually a 3-hp petrol start kerosene run engine operates the weeder. The engine power is transmitted to ground wheels through V belt-pulley and sprocket - chain mechanism. At the back of the machine a replaceable sweep blade is fixed (Fig. 19.2). Sweep blades of different width can be fitted to the machine depending on the row to row spacing of the crop. A tail wheel is provided at the rear to maintain the operating depth. The sweep blade can be raised or lowered so as to have the desired operating depth. The salient features of the unit are:

- Useful for weeding between rows of crops like tapioca, cotton, sugarcane, maize, tomato and pulses whose rows spacing is more than 45 cm
- Can also be used for weeding and intercultural operations in orchards, coconut and areca nut plantations.

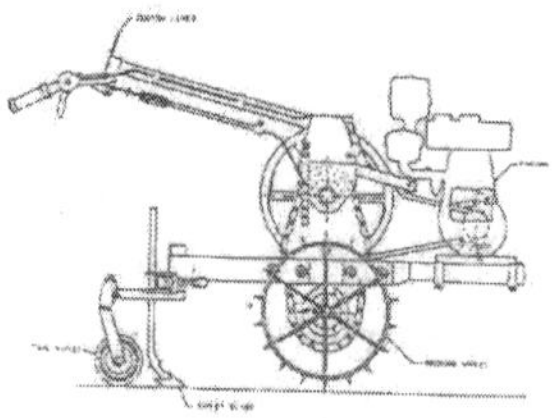

Fig. 19.2 Engine operated weeder

SWEEP

It is an intercultural implement for removing shallow rooted weeds in between rows. The sweep consists of V shaped shovels with bevel edged wings. The shovels are held by the tynes fixed to a frame by means of counter sunk bolts and nuts. When the sweep is used for secondary tillage, five or six tynes may be clamped with the shovels in line having no gap in between them (Fig. 19.3). By just skimming under the soil at a shallow depth of 2 to 3 cm, the sweep breaks the capillary in the soil pores and provides soil mulch. When the sweep is used for intercultural operations, the space between the shovels is adjusted to suit the row spacing of the crop and with different sizes of blades. The area covered by sweep is 1.75 to 2.5 ha/day. The salient features of the unit are:

- Suitable for all row crops and soils; provides soil mulch and conserves soil moisture
- Suitable for inter cultural operations.

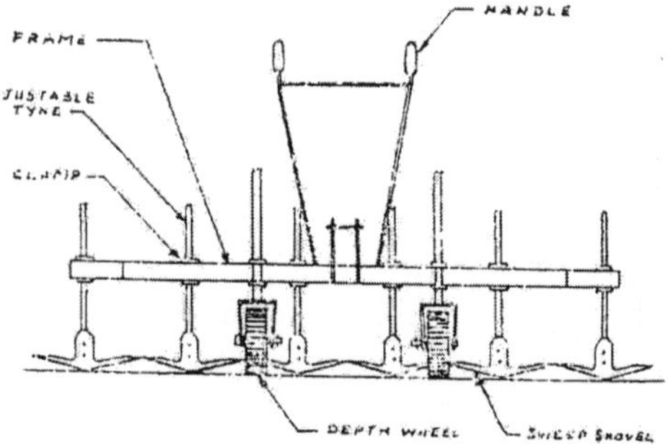

Fig. 19.3 Bullock operated sweeps

JUNIOR HOE

It is intercultural equipment used primarily for weeding in between the rows of standing crops. It consists of reversible shovels with curved tynes attached to framework with hinge arrangement. A handle and beam are fixed to the framework for guiding and attaching the unit to the yoke (Fig. 19.4). The spacing between the shovel can be adjusted according to the row spacing of the crop. The area covered by the junior hoe is about 1.5 ha per day.

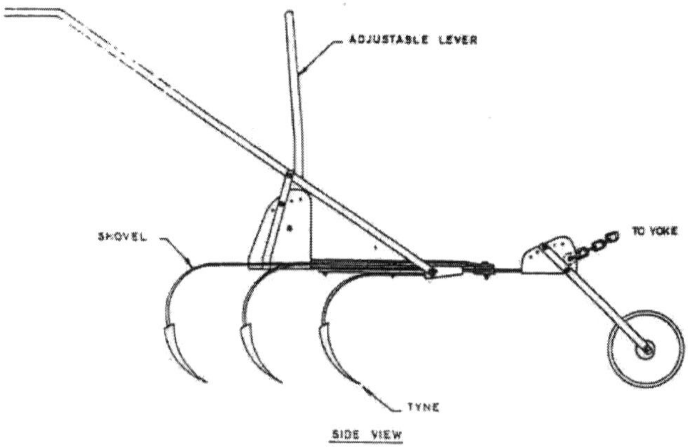

Fig. 19.4 Junior hoe

Intercultural operations in horticultural crops

Sl No.	Crop	Intercultural operations
1	Beet	Weed Control, Thinning
2	Bitter Gourd	Weeding
3	Brinjal	Weeding By Shallow Cultivation, Earthing up
4	Carrot	Weeding and hoeing, Earthing up
5	Cauliflower	Weeding, hoeing, earthing up, Blanching
6	Cabbage	Weeding, hoeing, earthing up
7	Okra	Weeding
8	Onion	Weeding
9	Peas	Weeding, Vine Support
10	Radish	Weeding and hoeing, Earthing up
11	Tomato	Weed Control ,shallow Cultivation for aeration, earthing up, mulching
12	Mango	Training and pruning ,hoeing(crust breaking),Weeding and Ploughing

13	Citrus	Weed Control, Intercropping, Training and Pruning
14	Papaya	Thinning, Weed Control, Intercropping

Tools used in intercultural operations

Sl. No.	Name of Tool/implements Used for inter cultivation	Intercultural operation
1	Hoe	Weeding, thinning, crust breaking, soil aeration.
2	Cultivator. Fork	Weeding, loosening the soil for aeration
3	Ridge Plough, Spade	earthing up
4	Poles	Wine support
5	Knife	Training
6	Rope Saws, Pole Pruner & Lopper, Folding Pruner	Pruning

Other hand tools used in horticulture

S. No.	**Name of hand tool**	**Usage**
1	Spade	Digging, trenching and removing soil
2	Fork	Used after the soil is turned by the spade for breaking clods and Weeding
3	Border fork	For weeding in borders
4	Potato fork	For digging out potatoes, for clearing compost and rubbish
5	Garden rake-	To break up the clods, leave to a fine tilth. It Can also be used for gathering stones, weeds and hedge clippings
6	Lawn rake	Used on lawns to rake out the dead grass and collect leaves
	The hay rake	It is used for raking together grass and leaves
	Draw hoe	It chops through the roots of the weeds, especially tap roots, and can also be used for making drills for seed sowing and for singling
	Dutch hoe	It is used for Weeding. It is given with a pushing action as the operator walks
	Backwards	
	Triangular headed	

Hoe	Used with a pulling action for making a Shallow drill	
	Trowels	They are small scoop shaped tools used for making small holes for planting plants
	Garden line	It is a very important horticultural tool used for setting out straight lines. It is used when seed sowing, trenching, lawn edging, transplanting and for many other garden operations
	Hand shears	Used for cutting grass and hedges
	Edging shares	Used for cutting lawn edges and because of the length of the handles they can be used when standing upright.

20

HARVESTING EQUIPMENTS

It is the operation of cutting, picking, plucking digging or a combination of these operations for removing the crop from under the ground or above the ground or removing the useful part or fruits from plants.

Harvesting action can be done by four ways:

(1) Slicing action with a sharp tool.

(2) Tearing action with a rough serrated edge.

(3) High velocity single element impact with sharp or dull edge.

(4) Two elements scissors type action.

The harvesting involves slicing and tearing action. The harvesting can be done either by manually operated tool or animal drawn equipment or mechanically operated machine.

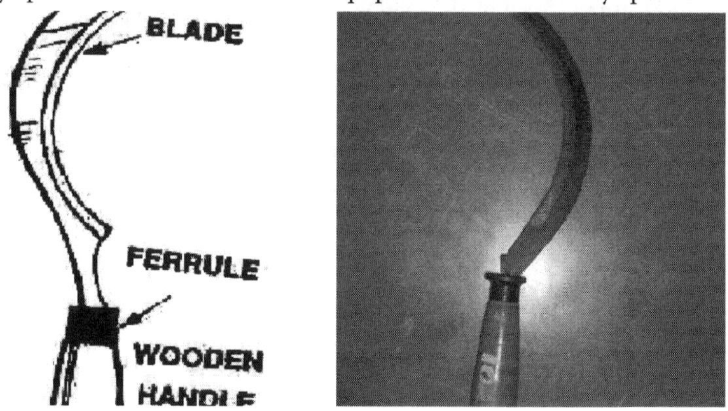

Fig. 20.1 Sickle

Sickle: Sickle is a simple harvesting tool. It is used for harvesting crops and cutting other vegetations. The sickle mainly consists of a metallic blade and a wooden handle.

Blade is the main metallic part of the sickle. The blade should be made of carbon steel. The blade is made in curved shape. There are two types of metallic cutting blade such as plain and serrated. The teeth of serrated sickle are made sharp for efficient working in the field. The handle of the sickle is made of well seasoned wood. The forged end of the blade for fixing the handle is called tang. Protective metallic bush fitted at the junction of the blade and the handle to keep the tang tight in the handle is called ferrule. The parts of sickle are shown in Figure. Harvesting by sickle is a very slow and labour consuming device.

POTATO HARVESTER

On large size farms where potato cultivation is practiced on commercial scale, animal drawn cultivators or country ploughs were used for harvesting the potatoes. However, the implement consisting of a ridge bottom and a replaceable bar point share has been successfully employed for this purpose. The potato soil mass while moving on the rods provided at the back of the share, gets separated. The potatoes are then dropped on the ground which is later collected manually. The same implement has been modified and can be hitched by a power tiller also. The animal drawn groundnut digger can be successfully used for digging of potatoes.

A potato digger suitably designed to be operated as a mounted machine by a tractor. The potato harvester is used extensively for harvesting potatoes in Northern States of India. The machine uses conventional digging blade and a shaker conveyor made of steel rods. As the tractor with the machine moves over the field, the soil and potatoes are dug out and lifted on the conveyor chain. The soil is shaken out by agitation where as the potatoes are collected in bags.

The tractor operated potato digger consists of a three point hitch frame, main frame, and two column frames with blade. Blade is fixed to with two detachable ridge opening devices. On the sides of the columns, two furrow guides are provided for stability of the digger. Two replaceable high carbon steel shovels are fitted in front of the blade for easy penetration in dry field conditions and to prevent it from wear.

The tractor powered potato digger can be conveniently divided into four broad categories:

1. Blade type diggers,
2. Single row spinner digger type machines,
3. Single or double row elevator diggers,
4. Oscillating / vibrator diggers.

The most common among the four designs as stated above is the elevator type digger. The commercial machines are of both single row and two row types. The potato harvester is shown in plate: . The machine consists of a shovel type digger and endless elevator to complete the total operation.

Plate : 20.2 a b Tractor operated Potato harvester

(Source : http://img.esuppliersindia.com & http://www.huaxinmachine.com)

TAPIOCA HARVESTER

The tapioca harvester is tractor operated implement. It consists of a main frame with three point hitch system, shank with depth adjustment and digging blades with pegs. Provision is made to for adding dead weight on the main frame. The digging blade is trapezoidal in shape and five pegs are fitted in front of the blade for easy penetration into the soil and reduce the draft requirement of the unit. Before digging operation the top portion of the plant has to be removed. When the implement is pulled forward, tapioca tubers are uprooted.

The salient features of the are : harvesting efficiency is about 98%, can be easily attachable to any make of 35 hp tractor, reduction in harvesting cost compared to manual harvesting, tubers are not left as undug in the field and additional irrigation for manual digging is eliminated.

TURMERIC HARVESTER

Turmeric harvester is a power tiller operated unit. It consists of a blade with three bar points for easy penetration into the soil. To the rear end of the blade six slats spaced at 50 mm apart are hinged at both ends. The oscillating motion for the slats is obtained through eccentric provided on either side of the unit. The power is transmitted from the clutch pulley of the power tiller to a reduction gearbox mounted near the hitch bracket assembly of the power tiller. From the gear box, the power is transmitted to the shaft of the turmeric digger unit through V belt transmission.

FRUIT HARVESTER

It is estimated that 70% of the potential of lasting quality of many fruits is predetermined at harvest; post harvest factors accounting for the remaining 30%. Certain harvest factors such as time of harvest, or harvest maturity and method of harvest also influence the keeping quality of perishables like fruits, besides their mechanical properties. In India, manual harvesting is in vogue using ladders, poles, 'dhoti', chippers, etc. The large size of fruit trees poses a great problem in manual harvesting resulting in more damage.

Improved harvesting methods would help not only to reduce these losses but also improve harvesting to 40 to 65 kg/ha compared to 15 to 25 kg/ha with traditional methods.

There are two types of manually operated fruit harvesters such a fixed height type with wooden head and metal cutting blade and adjustable height type with metal head and metal cutting blade.

1. *Fixed height with wooden head and metal blade harvester*

This type of harvester has a light weight cutting mechanism made of a light bamboo pole. The components of the harvester include, a wooden head, a metal blade, a spring, a fruit collecting net, a wooden pole, a hand lever and nylon rope. The cutting mechanism of the harvester consists of the wooden head, the cutting blade and the retaining spring. The wooden head of 220 X 115 X 20 mm is stationary and acts as a support against the cutting blade when a fruit stem comes between the metal blade of 195 X 58 X 6 mm in size and the head. The failure of the stem takes place due to the cutting action of the sharp blade against the head. The blade is given proper sharpness and heat treatment to prevent fast wear. The blade is always held in open position by means of a soft spring of 64 mm in length and 90 mm in diameter. The blade moves against the tension of the spring while cutting is taking place. Immediately after cutting, the blade is pulled back to its original open position by the spring which is connected between the blade and the pole. A nylon net is fixed around a ring of 300 mm diameter made of 8 mm steel rod, just below the cutting mechanism to collect the cut fruit, thus a avoiding any damage. The net depth is kept at about 270 mm. The entire cutting mechanism was fitted to a bamboo pole of 12 mm diameter and 3.5 m length. At about 380 mm from the bottom end, a hand lever is fitted to the pole. The metal cutting blade on the top and the hand lever are connected by a nylon rope of 3.2 m length. By pushing the hand lever down, the blade moves against the head and cutting takes place.

2. Adjustable height with metal head and metal blade harvester

In this type of harvester, the cutting mechanism is similar to the fixed height type harvester. The variations are only in head and pole length to achieve increased adjustable height up to 5.4m. An additional base plate is provided at the bottom of the pole. The cutting head was fabricated out of angle iron of 191 X 115 X 6 mm size. The pole is composed of two G.I. pipes of 12 mm and 18 mm sizes. One can be inserted in to the other. This telescopic arrangement of poles is helpful in varying the height. The pole can be fixed at any desired height by a clamp arrangement, thus making it flexible, adjustable and not fixed as in earlier harvester.

The cutting mechanism is arranged at the top of a 12 mm pipe by means of a coupling. This facilitates dismantling and assembly of the cutting mechanism from the pole. Provision of base plate at the bottom of the pole with U-clamp arrangement helps to freely move pole in all directions to meet fruit orientation. This facilitates the operation of the harvester although it is heavier than the fixed type harvester discussed earlier.

HOISTS

A hoist (Plate : 20.2) is a device used for lifting or lowering a load by means of a drum or lift-wheel around which rope or chain wraps. It may be manually operated, electrically or pneumatically driven and may use chain, fiber or wire rope as its lifting medium. The load is attached to the hoist by means of a lifting hook.

Fig: 20.3 Petrol engine operated Hoist

21

MODERN TRACTORS

A tractor is a vehicle specifically designed to deliver a high tractive effort (or torque) at slow speeds, for the purposes of hauling a trailer or machinery used in agriculture or construction. Most commonly, the term is used to describe a farm vehicle that provides the power and traction to mechanize agricultural tasks, especially (and originally) tillage but nowadays a great variety of tasks. Agricultural implements may be towed behind or mounted on the tractor, and the tractor may also provide a source of power if the implement is mechanised.

The word tractor was taken from Latin, being the agent noun oftrahere "to pull". The first recorded use of the word meaning "an engine or vehicle for pulling wagons or ploughs" occurred in 1901, displacing the earlier term traction engine

HISTORY

The first powered farm implements in the early 19th century were portable engines - steam engines on wheels that could be used to drive mechanical farm machinery by way of a flexible belt. Around 1850, the first traction engines were developed from these, and were widely adopted for agricultural use. The first tractors were steam-powered plowing engines. They were used in pairs, placed on either side of a field to haul a plow back and forth between them using a wire cable. Where soil conditions permitted (as in the United States) steam tractors were used to direct-haul plows, but in the UK and elsewhere plowing engines were used for cable-hauled plowing instead. Steam-powered agricultural engines remained in use well into the 20th century until reliable internal combustion engines had been developed.

In 1892, John Froelich invented and built the first gasoline/petrol-powered tractor in Clayton County, Iowa, USA. After receiving a patent Froelich started up the Waterloo Gasoline Engine Company, investing all of his assets which by 1895, all would be lost and his business resigned to become a failure.

After graduating from the University of Wisconsin, Charles W. Hart and Charles H. Parr developed a two-cylinder gasoline engine and set up their business in Charles City, Iowa. In 1903 the firm built fifteen "tractors". A term with Latin roots coined by Hart and Parr and a combination of the words traction and power. The 14,000 pound #3 is the oldest surviving internal combustion engine tractor in the United States and is on display at the Smithsonian National Museum of American History in Washington D.C. The two-cylinder engine has a unique hit-and-miss firing cycle that produced 30 horsepower at the belt and 18 at the drawbar.

In Britain, the first recorded tractor sale was the oil-burning Hornsby-Ackroyd Patent Safety Oil Traction engine, in 1897. However, the first commercially successful design was Dan Albone's three-wheel Ivel tractor of 1902. In 1908, the Saunderson Tractor and Implement Co. of Bedford introduced a four-wheel design, and went on to become the largest tractor manufacturer outside the U.S. at that time.

While unpopular at first, these gasoline-powered machines began to catch on in the 1910s when they became smaller and more affordable. Henry Ford introduced the Fordson, the first mass-produced tractor in 1917. They were built in the U.S., Ireland, England and Russia and by 1923, Fordson had 77% of the U.S. market. The Fordson dispensed with a frame, using the strength of the engine block to hold the machine together. By the 1920s, tractors with a gasoline-powered internal combustion engine had become the norm.

SAFETY

Agriculture in the United States is one of the most hazardous industries, only surpassed by mining and construction. No other farm machine is so identified with the hazards of production agriculture as the tractor. Tractor-related injuries account for approximately 32% of the fatalities and 6% of the non-fatal injuries in agriculture. Over 50% is attributed to tractor overturns.

The roll over protection structure (ROPS) and seat belt, when worn, are the two most important safety devices to protect operators from death during tractor overturns.

Modern tractors have rollover protection systems (ROPS) to prevent an operator from being crushed if the tractor overturns. It is important to remember that the ROPS does not prevent tractor overturns. Rather, it prevents the operator from being crushed during an overturn. This is especially important in open-air tractors, where the ROPS is a steel beam that extends above the operator's seat. For tractors with operator cabs, the ROPS is part of the frame of the cab. A ROPS with enclosed cab further reduces the likelihood of serious injury because the operator is protected by the sides and windows of the cab.

ROPS were first required by legislation in Sweden in 1959. Before ROPS were required, some farmers died when their tractors rolled on top of them. Row-crop tractors, before ROPS, were particularly dangerous because of their 'tricycle' design with the two front wheels spaced close together and angled inward toward the ground. Some farmers were killed by rollovers while operating tractors along steep slopes. Others have been killed while attempting to tow or pull an excessive load from above axle height, or when cold weather caused the tires to freeze to the ground, in both cases causing the tractor to pivot around the rear axle.

For the ROPS to work as designed, the operator must stay within the protective frame of the ROPS. This means the operator must wear the seat belt. Not wearing the seat belt may defeat the primary purpose of the ROPS.

APPLICATIONS AND VARIATIONS

Farm tractor applications

The most common use of the term "tractor" is for the vehicles used on farms. The farm tractor is used for pulling or pushing agricultural machinery or trailers, for plowing, tilling, disking, harrowing, planting, and similar tasks.

A variety of specialty farm tractors have been developed for particular uses. These include "row crop" tractors with adjustable tread width to allow the tractor to pass down rows of corn, tomatoes or other crops without crushing the plants, "wheatland" or "standard" tractors with non-adjustable fixed wheels and a lower center of gravity for plowing and other heavy field work for broadcast crops, and "high crop" tractors with adjustable tread and increased ground clearance, often used in the cultivation of cotton and other high-growing row crop plant operations, and "utility tractors", typically smaller tractors with a low center of gravity and short turning radius, used for general purposes around the farmstead. Many utility tractors are used for non-farm grading, landscape maintenance and excavation purposes, particularly with loaders, backhoes, pallet forks and similar devices. Small garden or lawn tractors designed for suburban and semi-rural gardening and landscape maintenance also exist in a variety of configurations.

Some farm-type tractors are found elsewhere than on farms: with large universities' gardening departments, in public parks, or for highway workman use with blowtorch cylinders strapped to its sides and a pneumatic drill air compressor permanently fastened over its power take-off. These are often fitted with grass (turf) tyres which are less damaging to soft surfaces than agricultural tires.

Precision agriculture

Space technology has been incorporated into agriculture in the form of GPS devices, and robust on-board computers installed as optional features on farm tractors. These technologies are used in modern, precision farming techniques. The spin-offs from the space race have actually facilitated automation in plowing and the use of autosteer systems drone on tractors that are manned but only steered at the end of a row, the idea being to neither overlap and use more fuel nor leave streaks when performing jobs such as cultivating.

Engineering tractors

The durability and engine power of tractors made them very suitable for engineering tasks. Tractors can be fitted with engineering tools such as dozer blade, bucket, hoe, ripper, and so on. The most common attachments for the front of a tractor are dozer blade or a bucket. When attached with engineering tools the tractor is called an engineering vehicle.

A bulldozer is a track-type tractor attached with blade in the front and a rope-winch behind. Bulldozers are very powerful tractors and have excellent ground-hold, as their main tasks are to push or drag things.

Bulldozers have been further modified over time to evolve into new machines which are capable of working in ways that the original bulldozer can not. One example is that loader tractors were created by removing the blade and substituting a large volume bucket and hydraulic arms which can raise and lower the bucket, thus making it useful for scooping up earth, rock and similar loose material to load it into trucks.

A front-loader or loader is a tractor with an engineering tool which consists of two hydraulic powered arms on either side of the front engine compartment and a tilting implement. This is usually a wide open box called a bucket but other common attachments are a pallet fork and a bale grappler.

Other modifications to the original bulldozer include making the machine smaller to let it operate in small work areas where movement is limited. There are also tiny wheeled loaders, officially called Skid-steer loaders but nicknamed "Bobcat" after the original manufacturer, which are particularly suited for small excavation projects in confined areas.

The word 'tractor' originated from the Latin word 'trahere', meaning 'pull'. It is an engine or vehicle for pulling wagons or plows. Today, tractors are used for drawing in, towing or pulling objects that are extremely hard to move. One commonly sees a tractor on farms used to push agricultural machinery or trailers that plow or harrow fields. The first tractors from the 1800s and early 1900s, were powered by steam engines. These tractors were phased out due to the instability of the steam engine that caused explosions, or trapped the driver in a belt driven attachment. The successors were built with an internal combustion engine. Technology has found its way into modernizing the tractor. Nowadays, one finds Global Positioning Systems (GPS) and on-board computers attached to farm tractors. With the use of advance technology, corporate-scale farms are using a combination of unmanned tractors and manually driven vehicles to plow the fields.

A tractor is a vehicle specifically designed to deliver a high tractive effort at slow speeds, for the purposes of hauling a trailer or machinery used in agriculture or construction.

Most commonly, the term is used to describe the distinctive farm vehicle. Agricultural implements may be towed behind or mounted on the tractor, and the tractor may also provide a source of power if the implement is mechanized. Another common use of the term is for the power unit of a semi-trailer truck. The first tractors, powered by steam engines, were phased out followed by the internal combustion engine.

Modern tractors are built with a rollover protection system (ROPS) that protects the operator from being crushed in the event that the vehicle rolls over. It was the New Zealand legislation that first required the ROPS to be built in to tractors in the 1960s. Before this system was required, many farmers were killed in accidents when the tractors rolled on top of them, crushing them in the process. Usually these accidents happen when tractors were driven along steep

HISTORY OF TRACTORS

In 1892, John Froelich built the first practical gasoline-powered tractor in Clayton County, Iowa. Only two were sold, and it was not until 1911, when the Twin City Traction Engine Company developed the design, that it became successful. The classic farm tractor is a simple open vehicle, with two very large driving wheels on an axle below and slightly behind a single seat (the seat and steering wheel consequently are in the center), and the engine in front of the driver, with two steerable wheels below the engine compartment.

This basic design has remained unchanged for a number of years, but enclosed cabs are fitted on almost all modern models, for reasons of operator safety and comfort. Originally, plows and other equipment were connected via a draw-bar, or a proprietary connecting system; prior to Harry Ferguson patenting the three-point hitch.

Recently, Bobcat's patent on its front loader connection has expired; and compact tractors are now being outfitted with quick-connect attachments for their front-end loaders.

There are also lawn tractors. Cub Cadet, Husqvarna, John Deere, Massey Ferguson and Toro are some of the better-known brands

Uses of Tractors

Farm implements can be attached to the rear of the tractor by either a drawbar or a three-point hitch. The three-point hitch was invented by Harry Ferguson and has been standard since the 1960s. Equipment attached to the three-point hitch can be raised or lowered hydraulically with a control lever. The equipment attached to the three-point hitch is usually completely supported by the tractor. Another way to attach an implement is via a Quick Hitch, which is attached to the three-point hitch. This enables a single person to attach an implement quicker and put the person in less danger when attaching the implement.

In some farm-type tractor, blowtorch cylinders strapped to its sides are used with a pneumatic drill air compressor permanently fastened over its power take-off. These are often fitted with grass (turf) tyres which are less damaging to soft surfaces than agricultural tires. Supposedly industrial bar tires are less damaging to lawns and soft surfaces than agricultural tires, but provide similar traction, and have the benefit of being self-cleaning. Often, these can be seen on road construction backhoes. Equipment costs account for about 20 percent of the total costs of production. The cost for depreciation, interest and insurance can range from $25 to $60 per acre depending on the crop and how intensively the equipment is used.

Operation

Modern farm tractors usually have five foot-pedals for the operator on the floor of the tractor. The pedal on the left is the clutch. The operator presses on this pedal to disengage the transmission for either shifting gears or stopping the tractor. Two of the pedals on the right are the brakes. The left brake pedal stops the left rear wheel and the right brake pedal does the same with the right side. This independent left and right wheel braking augments the steering of the tractor when only the two rear wheels are driven. This is usually done when it is necessary to make a tight turn. The split brake pedal is also used in mud or soft dirt to control a tire that spins due to loss of traction. The operator presses both pedals together to stop the tractor. For tractors with additional front-wheel drive, this operation often engages the 4-wheel locking differential to help stop the tractor when travelling at road speeds.

A fifth pedal just in front of the seat operates the rear differential lock (diff. lock) which prevents wheelslip. The differential lock allows the outside wheel to travel faster than the inside one during a turn. However, in traction conditions on a soft surface, the same mechanism could allow one wheel to slip, thus preventing traction to the other wheel. The diff. lock overrides this, causing both wheels to supply equal traction. Care must be taken to unlock the differential, usually by hitting the pedal a second time, before turning, since the tractor cannot perform a turn with the diff. lock engaged.

Useful Steps in Operation of Tractors

- The tractor should be proceeded slowly and carefully
- Ditches, gullies and steep embankments should be avoided as far as possible
- Be aware that a towed trailer or heavy implement can easily upset a tractor's balance and it can overturn the tractor causing hazards
- Accelerate gradually against gravity's resistance. Tractors tend to flip over backwards while going uphill, especially with a rear implement or when towing
- Small rocks, holes and bumps should be covered carefully. They can quickly disturb a tractor's equilibrium

- If possible, always travel directly up or down a hill, never across the slope
- If cross slope operation is necessary, always turn down the hill with the slope, never uphill into the slope
- While transporting a load uphill, always attach the load to the front of the tractor and back up the hill
- When towing a load downhill, use a low gear and let the tractor's engine act as a brake.

TYPES OF TRACTORS

Compact Utility Tractor

As the name suggests these tractors are modified and assembled to serve various purposes. CUT, as it is commonly called is a smaller version of an agricultural tractor but designed primarily for landscaping and estate management type tasks rather than for planting and harvesting on a commercial scale. Typical CUTs range in from 20 to 50 horsepower (15-37 kW) with available power take off (PTO) horsepower ranging from 15 to 45 hp (11-34 kW). CUTs are often equipped with both a mid-mounted PTO and a standard rear PTO, especially those below 40 horsepower (30 kW). The mid-mount PTO shaft typically rotates at/near 2000 rpms and is typically used to power such implements as mid-mount finish mower, a front mounted snow blower or front mounted rotary broom. The rear PTO is standardized at 540 rpms for the North American markets, but in some parts of the world, a dual 540/1000 rpm PTO is standard and implements are available for either standard in those markets. One of the most common attachment for a Compact Utility Tractor is the front end loader or FEL. All modern CUTs feature a government mandated roll over protection structure (ROPS) just like agricultural tractors. Although less common, compact backhoes are often attached to compact utility tractors.

Farm Tractor

The most common use of the term tractor is for the vehicles used on farms. The farm tractor is used for pulling agricultural machinery or trailers, for plowing, harrowing and similar tasks. The classic farm tractor is a simple open vehicle with two very large driving wheels on an axle below and slightly behind a single seat (the seat and steering wheel consequently are in the center) and the engine in front of the driver with two steerable wheels below the engine compartment. This basic design has remained unchanged for a number of years, but now enclosed cabs are available for many models of farm tractor. There are usually four foot-pedals, for the operator, on the floor of a tractor. The pedal on the left is the clutch. The operator presses on this pedal to disengage the transmission for either shifting gears or stopping the tractor. Two of the pedals on the right are the brakes. The left brake pedal stops the left rear wheel and the right brake pedal does the same with the right side. This independent left and right wheel braking augments the steering of the tractor when only the two rear wheels are driven. This is usually done when it is necessary to make a tight turn.

Engineering Tractor

These tractors are fitted with engineering tools such as bucket, hoe, ripper, dozer blade, and so on. The most common attachments for the front of a tractor are the dozer blade or a bucket. Attached with engineering tools, the tractor is called an engineering vehicle and it is capable of performing complex processes. A bulldozer is one of this kinds of tracked-type tractor attached with blade in the front and a rope-winch behind. Bulldozers are very powerful tractors and have excellent ground-hold, as their main tasks are to push or drag things. Bulldozers have been further modified over time to evolve into new machines which are capable of working in ways that the original bulldozer cannot. One example is that loader tractors are created by removing the blade and substituting a large volume bucket and hydraulic arms which can raise and lower the bucket, thus making it useful for scooping up earth, rock and similar loose material to load it into trucks.

FRONT LOADER

A front-loader or loader is a tractor with an engineering tool which consists of two hydraulic powered arms on either side of the front engine compartment and a tilting implement. This is usually a wide open box called a bucket but other common attachments are a pallet fork and a bale grappler. Other modifications to the original bulldozer include making the machine smaller to let it operate in small work areas where movement is limited. There are also tiny wheeled loaders, officially called skid-steer loaders but nicknamed "Bobcat" after the original manufacturer. These types of tractors are particularly suited for small excavation projects in confined areas.

Backhoe Loader

The Backhoe Loader is a special purpose tractor used mainly in the construction industry. As the name implies, it has a loader assembly on the front and a backhoe on the back. When both the loader and the backhoe are permanently attached, it is almost a huge crane. When the backhoe is permanently attached, the machine usually has a seat that can swivel to the rear to face the hoe controls. Removable backhoe attachments almost always have a separate seat on the attachment. Backhoe-loaders are very common and can be used for a wide variety of tasks like construction, small demolitions, light transportation of building materials, powering building equipment, digging holes, breaking asphalt and paving roads. Some buckets have a retractable bottom, enabling them to empty their load more quickly and efficiently. Buckets with retractable bottoms are also often used for grading and scratching off sand. The front assembly may be a removable attachment or permanently mounted. Often the bucket can be replaced with other devices or tools.

Garden Tractor

These tractors are small in size and used mainly for domestic plowing and gardening. Garden Tractors have stronger frames, axles and transmissions. Garden Tractors are generally capable of mounting other implements such as harrows, cultivators, sweepers, rollers and dozer-blades. Garden Tractors (also called Mini Tractors) are small, light and simple tractors designed for use in domestic gardens. These are usually designed primarily for cutting grass, being fitted with horizontal rotary cutting decks like ride-on mowers. These type of tractors generally have a horizontally-mounted engine with a belt-drive to a transaxle-type transmission (usually of 4- or 5-speeds, although some my also have two-speed reduction gearboxes or hydraulic gearboxes).

EPA TRACTOR

An EPA tractor is made just like an automobile, truck or lorry, without the passenger space behind the front seats, equipped with two gearboxes in a row to give the speed of loading heavy weights..

Road Tractor

They are heavy-duty vehicles with large engines and several axles. The majority of these tractors are designed to pull long semi-trailers, most often to transport freight of some kind over a significant distance, and is connected to the trailer with a fifth wheel coupling.

Locomotive Tractor

Amalgamating machines, electrical generators, controls and devices that comprise the traction component of railway vehicles, these tractors are built to move heavy vehicles. They are used mainly for rail car movers.

TRACTOR HAZARDS AND SAFETY

Deaths involving tractors contribute to agriculture having one of the highest on-the-job annual death rates -- 20.3 deaths for 100,000 workers nationally in the United States of America. In a 1999 study conducted by the Deep-South Center for Agricultural Health and Safety, it was found that, on average, 21 adults died each year in Florida agriculture in recent years. At least 74 of those deaths involved machinery, including tractors overturning and crushing their drivers. Across the nation, approximately 36 per cent of the agricultural work fatalities in 1997 involved a tractor, and 19 percent involved other machinery, according to the National Safety Council.

Safer tractor operations and the use of protective equipment could prevent many of the deaths and injuries involving tractors. Owners, managers, and tractor operators all play a major role in making tractor operations safer. Owners have responsibilities for providing a safer environment and safer equipment. Managers coordinate maintenance designed to ensure the safety of equipment, as well as safety policies and programs. The final responsibility for making the operation of tractors safer lies with the tractor operators themselves. It's often only when an incident occurs that people become focused on what might have prevented injury or loss of life. The following pages give guidance to agricultural employees on making tractor use safer.

Implementation of Safety Rules

- Dismounting load from a tractor when the engine is running is not allowed, unless the operation requires it
- Consumption of alcohol or drugs, including medications that may impair judgment, during work hours or for four hours before starting work should be prohibited
- Always use hearing protection
- Carbon monoxide hazards can be detected by signs of severe headeache. Check to see if the exhaust is in a location that exposes the operator to exhaust fumes. There have been instances of carbon monoxide exposure using an open-station tractor in the field
- If the tractor has a ROPS, securely fasten the seat belt. Do not use seat belts on tractors that don't have ROPS
- Where possible, avoid operating the tractor near ditches, embankments, and holes
- Reduce speed when turning and crossing slopes and on rough, slick, or muddy surfaces
- Stay off slopes that are too steep for safe operation
- Watch carefully for obstacles and other hazards in the tractor's path, especially at the end of rows, on roads, around trees and in other blind corners.

Tractor Sales and Farmers

Facing market saturation in the traditional markets of the north west (Punjab, Haryana, eastern Uttar Pradesh) tractors sales began a slow and slight decline. By 2002 sales went below 200,000. Manufacturers scrambled to push into eastern and southern India markets in an attempt to reverse the decline, and began exploring the potential for overseas markets. Sales remained in a slump, and added to the market saturation problems also came increased problems of "prestige" loan defaults, where farmers who were not financially able took tractors in moves to increase their families prestige. There are also reported increased misuse of these loans for buying either lifestyle goods, or for social functions. Government and private banks have both tightened their lending for this sector adding to the industry and farmers woes.

By 2004, a slight up tick in sales was witnessed once again. But by 2006 sales once again were down to 216,000 and now in 2007-08 have slid further to just over 200,000.

THE ROAD AHEAD

There has been recent trends in the development of tractors. Two new forms of modernized tractor is a matter of discussion and research among agricultural scientists all over the world, one a 'farmer's vehicle', a universal agricultural machine which could complete all operations in a single pass. The other is a single purpose 'tractor robot', an operator-free agricultural vehicle. The pros and cons of each option are enumerated and requirements for the development of a tractor-robot are in discussion. The tractor-robot has a system of cameras for a round 'vision', a satellite positioning system, movement sensors, control computer and automated motor controls. A prototype of a tractor-robot is presented and assessed; it is equipped with a self-locking cab which can be opened manually if needed, and a standard three-point PTO (power take off) for implements. Tractors are an indispensable agricultural machinery which is all set to bring about revolutionary changes in the world agricultural production.

The tractor as a modern agricultural tool

The modern tractor has evolved over the years. Powered farm implements in the nineteenth century where steam engines. These were used to drive farm machinery mechanically. This was the birth of modern day tractors and was used for agricultural purposes. This was later followed by internal combustion gasoline or petrol powered tractors. At first the gasoline powered tractors were unpopular until the Fordson brand, powered by internal combustion engine was introduced and became a hit.

Modern tractors are classified into two wheel drive and four wheel drives. Most of the tractors in the market are classical open vehicles with two huge rear wheels and small front wheels which are very key to steering its direction. The engine is strategically placed for high traction and balance. Caterpillars and crawlers became popular in the early thirties in heavy or wet soils due to their superior floatation and traction.

The basic use of a tractor is in the agricultural field though innovative manufacturers have come up with specific tractors for specific. The row crop tractor is specifically for tending crops from the time they sprout to harvest time and combines well with the cultivator. We also have the garden or mini tractors which are very light and used in small scale or personal gardens. Two wheel tractors are more or less like the garden counterparts and have a huge loyalty in the United States of America and beyond. Lastly and not the least we have the orchard tractors which are tailored to pass under orchard trees unperturbed.

All these innovations and tractor designed are geared towards precision agriculture and popularised by the modern lifestyles of reduction of manual labour by employing use of machines. All in all, it should be noted that use of machinery is not a bed of roses if safety is not checked properly. In the United States of America, agriculture is one of the most hazardous industries behind mining and construction. When looking for tractors to buy, the safety features are some of the aspects that will lead you to decide the kind of tractor fit for you.

The country inherited a stagnant agriculture at the time of Independence. The traditional tools and implements relied mostly on human and animal power and used a negligible amount of commercial energy. However, successive governments realized the importance of agriculture and initiatives were taken for the growth of this sector. Increased investment in irrigation infrastructure, expansion of credit, marketing, and processing facilities.

Till the 1950s, use of tractors for agriculture was very limited. Tractor manufacturing in India started in 1961 with aggregate capacity to manufacture 11 000 tractors. Joint efforts made by the government and private sector have led to steady increase in the level of mechanization over the years.

Given that rains are not always timely and evenly distributed, farmers prefer pump sets as a more reliable and assured source of irrigation; as a result, energization of pump sets have been increasing rapidly. As on 31 March 2004, 14.1 million pump sets had been energized. Maharashtra has the maximum number of energized pump sets (2.4 million), followed by Andhra Pradesh (2.3 million). Earlier, the average capacity of the pump sets was 3.68 kW and a pump set on an average consumed 6004 kWh of electricity in that year (Central Electricity Authority,2005). However, owing to insufficient electricity supplies, some farmers have also procured diesel pump sets as a standby. In the recent past, concerted efforts of the government has led to an introduction of biomass and solar photovoltaic based pumping systems

As a result of increased mechanization in agriculture, crop production and rural agro processing emerged as one of the major consumers of commercial energy. The share of mechanical and electrical power in agriculture increased from 40% in 1971/72 to 84% in 2003/04. The availability of farm power per unit area (kW/ha) has been considered as one of the parameters of expressing the level of mechanization. Power availability for carrying out various agricultural operations has increased from 0.3 kW/ha in 1971/72 to the tune of 1.4 kW/ha in 2003/04.

Connected load in the agriculture sector in 2004 was estimated to be 51.84 GW, the number of consumers being 12.8 million. The electricity consumption in agriculture during 2003/04 was 87 089 GWh (second highest)?24.13% of the total electricity consumption. There was an increase of 3.08% in the electricity sales to the agriculture sector in 2003/04 over 2002/03 (CEA 2005). Electricity consumption in agriculture sector has been increasing mainly because of greater irrigation demand for new crop varieties and subsidized electricity to this sector. Moreover, due importance is not given to proper selection, installation, operation, and maintenance of pumping sets, as a result of which they do not operate at the desired level of efficiency, leading to huge waste of energy.

Agriculture (plantation/food) consumed 7 123 thousand tonnes of HSD (high-speed diesel) in 2003/04, accounting for 19.2% of the total HSD consumption during the year. Consumption of LDO (light diesel oil) and furnace oil for plantation in 2003/04 was 44 000 and 243 000 tonnes, respectively, accounting for 2.7% of the total LDO and 2.9% of the total furnace oil consumed in the country. Consumption of furnace oil for transport (agriculture retail trade) in the agriculture sector was 94 thousand tonnes (Ministry of Power and Natural Gas 2004). However, it is difficult to assess the total diesel consumption for agriculture from the available data.

What are tractors for?

As promised, here is the thrilling second installment of our multi-part series, "What are tractors for?"

Last week, Siri wrote about what is surely the most common use of tractors in agriculture - turning a field with grass or other plants growing on it into a field of bare dirt. In most of today's agriculture, this means huge tractors pulling huge implements. On small farms like ours, we use many of the same implements to do this job - plows, chisel plows, discs - just smaller versions of them.

Once a field is prepared for planting, on our farm the next step is putting seeds or seedlings in the ground. On bigger farms than ours, farmers may plant only one or two crops once or twice a year. Market farms like ours, though, need to have a consistent supply of vegetables for months on end. You can't achieve this by planting acres of lettuce and kale just once in the spring. Instead, we plant smaller quantities, just a bed or two at a time, of a diverse mix of vegetables every week. This is where our smaller tractors, and especially our new 1950 Farmall Cub come into play. Our Cub does four jobs on the farm that modern tractors aren't as good at doing. It creates our beds, it marks rows for transplanting, it seeds, and it cultivates.

We make our beds on the farm by driving the Cub over recently tilled soil in as-straight-as-possible lines back and forth every 5 feet. Every bed on our farm is 5 feet wide by 125 feet long. Coincidentally, the Cub has 8 inch wide tires that are spaced 5 feet apart, center to center. This means, when we drive the tractor over freshly tilled soil, the tires create a depression where we can walk and leave a 4 foot wide raised bed between the tires for us to plant into. On our farm we set up flags every 5 feet on both ends of the bed for the tractor driver to aim at, ensuring that our beds are mostly straight and well spaced.

At the same time that the Cub is marking a bed, it might also be marking rows for transplanting. For ease of cultivation (which I'll discuss shortly), all the rows of vegetables within every bed are spaced identically. That is to say, we plant a maximum of three rows per bed, each spaced 15 inches apart. The Cub marks rows to transplant into by setting wheels down on the bed-top that make a line in the soil the length of the bed. Once a bed is marked, we'll return with trays of seedlings to transplant into one, two, or all three rows depending on the plants' need for space.

The other way we get rows of vegetables is by using our tractor mounted seeder. This year we purchased three Jang Seeders, which are ganged together and mounted on the back of the Cub. By lowering the rear hydraulics on the tractor, the seeders are set on the ground. As the tractor drives forward, the ground turns the front wheel of the seeder. This wheel, in turn, turns a barrel inside the seed hopper which picks up one seed at a time and drops it into the ground. The seeder then covers the seeds and compresses the soil behind them.

The last thing the Cub does is cultivate. The word cultivate is a funny one, because it has two agricultural meanings. The meanings most people know are "to grow plants," or "to prepare land for farming." That is to say, all the plowing, discing, and tilling Siri talked about in the prior post. What we mean when we say cultivate is to kill weeds. That is to say, hand weeding, hoeing, and tractor cultivating with the Cub. Beneath the belly of the Cub are mounted a set of knives that can be raised and lowered with the tractor's hydraulics.

To cultivate with the tractor, we move these knives so that they disturb the ground in the space between the rows of vegetables, killing only the weeds. Tractor cultivation is very tricky, for if the tractor driver goes off course even a little it can mean killing lots of vegetables. The engine and seat of the cub are offset to the left so the driver can keep a close watch on the knives and the vegetables.

Most of today's farms use chemicals to kill weeds, and no company that I know of has manufactured a cultivating tractor since the 1980s. So on farms like ours, which don't use chemicals for weed control, you'll often find old fashioned tractors like our Farmall Cub or an Allis Chalmers G for killing weeds.

There are tons of other uses for tractors on a farm. Today, I pulled an ancient cedar root out of the field. Next week we'll use the front end loader on a tractor like a cherry picker to help us work on the very high roof of our newest greenhouse. One of the funny things about becoming immersed in a subject is how you forget how little you knew before you began. After Siri wrote the first post about tractors, a number of people have admitted to us that they didn't actually understand much of the process necessary to turn a grassy field into a field of vegetables. Moreover, it seems that many of the details about what we do on the farm are lost on our audience because most people aren't familiar with what farmers actually do or the terms we use to describe our work. We ourselves didn't know much about the technicalities of growing food before we started farming. The role tractors play is but one small part of the whole process. But now, at least, you'll understand that part.

- One of the most important technological changes in agriculture was the transition to the modern tractor.

- The transition from horses and mules to modern farm machinery for tilling the soil and harvesting the crop did not begin until the 1920s.
- Tractors are big machines. They can do many different kinds of work. Normally farmers use them to pull plows and to work other farm machines. Farm tractors are designed for use in flat fields.
- A tractor is a machine that pulls or pushes a tool over land. Without tractor power and purpose-built implements, it's unlikely that anyone would look forward to work around the modern farm whether it is a small scale or lifestyle farm or a large commercial farm.
- Modern farm tractors are effective and up-to-date agricultural power units. Changes have helped make them more efficient, safe, convenient, versatile and powerful.
- Modern farm tractors may be divided into two general classes: the general purpose type; and the standard tread type.
- The general purpose type of tractor, as its name implies, furnishes power for practically all farm work.
- Not only does it perform all of the drawbar, belt and power shaft jobs, but, in addition puts speed and economy into the planting and cultivating of row crops.
- New tractors can provide high levels of operator comfort, for example air-conditioning and radios in the cabin. These comfort features tend to lessen operator fatigue and therefore improve safety.
- Modern farm tractors are equipped with receivers which using signals from Global Positioning System (GPS) satellites, allow precision hands-free plowing, planting and harvesting. Automated seed planters are synchronized with actual tractor speed by ground speed radar.
- Modern tractors are equipped with roller protection system of ROPS, consisting mainly of a steel framework which if the operator is not thrown out, should prevent the weight fo the oveturned machine from crushing him.
- Throughout the industrialized world, mechanization has become a cornerstone of production agriculture and the tractor is the modern symbol of farm mechanization.

22

ECONOMICS OF FARMING

AGRICULTURAL MACHINERY

Agricultural machinery is machinery used in the operation of an agricultural area or farm.

Steam Power

Power for agricultural machinery was originally supplied by horses or other domesticated animals. With the invention of steam powercame the portable engine, and later the traction engine, a multipurpose, mobile energy source that was the ground-crawling cousin to the steam locomotive. Agricultural steam engines took over the heavy pulling work of horses, and were also equipped with a pulley that could power stationary machines via the use of a long belt. The steam-powered machines were low-powered by today's standards but, because of their size and their low gear ratios, they could provide a large drawbar pull. Their slow speed led farmers to comment that tractors had two speeds: "slow, and darn slow."

Internal combustion engines

The internal combustion engine; first the petrol engine, and later diesel engines; became the main source of power for the next generation of tractors. These engines also contributed to the development of the self-propelled, combined harvester and thresher, or combine harvester (also shortened to 'combine'). Instead of cutting the grain stalks and transporting them to a stationary threshing machine, these combines cut, threshed, and separated the grain while moving continuously through the field.

Combines might have taken the harvesting job away from tractors, but tractors still do the majority of work on a modern farm. They are used to pull implements-machines that till the ground, plant seed, and perform other tasks.

Tillage implements prepare the soil for planting by loosening the soil and killing weeds or competing plants. The best-known is the plow, the ancient implement that was upgraded in 1838 by John Deere. Plows are now used less frequently in the U.S. than formerly, with offset disks used instead to turn over the soil, and chisels used to gain the depth needed to retain moisture.

- The most common type of seeder is called a planter, and spaces seeds out equally in long rows, which are usually two to three feet apart. Some crops are planted by drills, which put out much more seed in rows less than a foot apart, blanketing the field with crops. Transplanters automate the task of transplanting seedlings to the field. With the widespread use of plastic mulch, plastic mulch layers, transplanters, and seeders lay down long rows of plastic, and plant through them automatically.
- After planting, other implements can be used to cultivate weeds from between rows, or to spread fertilizer and pesticides. Hay balers can be used to tightly package grass or alfalfa into a storable form for the winter months.
- Modern irrigation relies on machinery. Engines, pumps and other specialized gear provide waterquickly and in high volumes to large areas of land. Similar types of equipment can be used to deliver fertilizers and pesticides.
- Besides the tractor, other vehicles have been adapted for use in farming, including trucks, airplanes, and helicopters, such as for trans

Technology

Only with mass production being aided by modern technology and intensive marketing can the agriculturist exploit both the domestic market as well as the international market to the fullest extent. The volume of production depends not only on the capital investments and marketing strategies but also on the technical capacity used during the production and processing stage.

In fact, technology has come to play a very significant role even in marketing these days. Technology is absolutely critical to the agro industry be it at the primary (production), secondary (processing) or tertiary (marketing and packaging) stage. Experts have always suggested that agricultural technology could play a vital role in addressing the issues and concerns relating to the conservation and management of rural resources.

Ploughing is the first preparation for planting. The plough is primarily designed to prepare the ground for cultivation by turning it over, thus burying the weeds and loosening the earth. It is generally agreed by historians that the earliest implement used for cultivation was probably a crude pointed bent stick or tree branch which was used to stir the soil surface. In effect, a hand held hoe was used in which the user scratched at the earth to form a tilth where corn could be sown. Over a period of time, these hand held hoes soon developed into simple ploughs. These primitive ploughs were eventually pulled by animals like oxen, camels and even elephants. Animals enabled the land to be tilled more easily and faster; thus more food was produced. The credit for this innovation goes to the Egyptians. These ploughs had different modifications in different parts of the world.

During the period from the 1820s to the 1840s, several innovations occurred in plough production. The Breaking Plough, or Prairie Breaker, was a heavy wooden plough plated with iron strips to reduce friction. Prairie ploughs were heavy, weighing at least 125 pounds and requiring from three to seven yoke of oxen. Cutting only three inches into the soil, farmers could break eight acres a year. After a span of few years, ploughs containing a polished wrought iron moldboard and steel share were invented.

Harrow

After ploughing, other implements were used. The harrow was necessary to smoothen the soil in areas where the soil remained rough. It consists of a wooden or metal framework bearing metal disks, teeth, or sharp projecting points, called tines, which is dragged over plowed land to crush the clods of earth and level the soil. Harrows are also used to uproot weeds, aerate the soil, and cover seeds.

In the beginning the harrows were as simple as a tree branch but the harrow became more sophisticated after the Industrial Revolution. By the 1790s, two distinct types of harrows were in use: the square and the triangle, or "A" frame. The square harrow was used on old fields that were free of large obstructions, while the triangular frame was used on freshly ploughed fields. These models had wooden frames with wood or iron teeth.

In modern times, harrows are of varied types. Some are simply dragged behind a tractor or draft animal; some are suspended on wheels; many have levers to adjust the depth of the cut.

IMPACT OF THE INDUSTRIAL REVOLUTION ON AGRICULTURE

Industrial revolution brought about drastic changes in the farming process. Farmers discovered the crop rotation system that allowed them to forgo leaving up to half of the land unused or fallow between each planting. This phase witnessed the use of animal husbandry. The industrial revolution brought about an end to tedious handiwork and encouraged the invention and manufacture of other labor-saving farm implements and machinery. Few of the inventions include:

Seed Driller

Seed drill was an innovation that allowed seeds to be easily planted deep into the earth instead of on top where the majority were washed away or otherwise lost. The machine was pulled by horses and consisted of rotating drills or runners that planted seeds at a set depth.

Horse Hoe

It is horse-drawn machine which loosened the soil and killed weeds.

The first reapers cut the standing grain and, with a revolving reel, sweeping it onto a platform from which it was raked off into piles by a man walking alongside. The reaper could thus harvest more grain than five men using the earlier cradles.

The reaper was eventually replaced by the self-propelled combine, operated by one man, which cuts, gathers, threshes, and sacks the grain mechanically. The reaper was the first step in a transition from hand labor to the mechanized farming of today.

Threshing Machines

Prior to the threshing machines farmers used an implement called 'flail' to simply beat the grain with sticks or ropes to knock the seeds from the stalks. But this was a back-breaking work and was of low productivity. Threshing machines were designed for rapidly removing the husks from grain.

With improvements in design and efficiency, threshing machines became progressively more common and the hand flail was gradually consigned to history. The machines could be driven by wind or water power, or by horses, but the steam powered thresher became the most familiar sight. They were eventually replaced in the middle decades of the twentieth century by the combine harvester which both harvests and threshes the crop in the field in a single operation.

Tractor

Tractor is a vehicle particularly crafted to exert traction at slow speeds, for the purposes of hauling a trailer or machinery used in agriculture. The versatility of tractor is with respect to its attachments that it supports. The more the options for connecting attachments to the tractor, the higher is the cost. The most common tractor attachments include front end loaders, mowers, box blades, spreaders, tillers, plows, trailers and backhoes for plowing, tilling, disking, harrowing, planting, and similar tasks. Most tractor attachments are interchangeable so they can be used with different machines. The first tractors were steam-powered ploughing engines, followed by Gasoline Powered Tractors.

AGRICULTURAL MACHINERY INDUSTRY IN INDIA

The country witnessed unprecedented growth in agriculture which has helped India to graduate from hunger to self sufficiency in food grains by increasing the food grain production from 51 million tonnes to 208 million tonnes, with surplus for export. The technology back-up by agricultural scientists, in the form of "Green Revolution" combined with industrial growth, positive policy support, liberal public funding for agricultural research and development and dedicated work of farmers contributed to the phenomenal increase in agricultural, animal and fish production. Application of engineering in agriculture was equally appreciated by the farmers and to-day they feel proud to have improved machinery from Bakhars to rotavators, Persian wheel to drip and micro-sprinkler systems, cone-dibblers to pneumatic planters, sickles to combine harvesters, sieve to colour sorters, and, kolhus to solvent extraction plants, and hand mills to roller flour mills, etc. The farmers are not afraid of hot/cold desert and vagaries of weather as they have green houses and low tunnel plastic houses technology to grow crops in any place at any time of the year.

The growth in adoption of agricultural machinery in the country has be impossible due to their local manufacture. The manufacture of agricultural machinery in India is under taken by village artisans, tiny units, small scale industries, organized medium and large scale sector. Organized sectors manufacture sophisticated machinery such as tractors, engines, milling and dairying equipment. Traditional hand tools and bullock drawn implements are largely fabricated by village craftsmen and power operated machinery by small scale industries.

Historical Perspective of Manufacturing of Agricultural Machinery

Blacksmiths and carpenters have been the traditional fabricators of agricultural equipment in India. The early agricultural mechanization in India was greatly influenced by the technological development in England.

In 1889, Watts and Kaisar introduced ploughs, corn grinders and chaff cutters Cawnpore (now Kanpur) Experimental Farm in Uttar Pradesh. Sardar Joginder Singh (1897-1946), who was the Agriculture Minister in the Punjab Government (1926-37), introduced the steam tractors in India in 1914 for reclamation of waste land and eradication of 'Kans'. Horse drawn and steam tractor operated implements were imported during the latter part of the 19th century. The horse drawn equipment imported from England were not suitable for bullocks and he-buffaloes used in India and thus, were suitably modified by small scale manufacturers to suit Indian draught animals. With the establishment of Allahabad Agricultural Institute, Allahabad in 1942, the development activities in agricultural machinery accelerated and as a result bullock drawn Meston, Shabash and Wah-Wah ploughs were introduced in Uttar Pradesh, manufactured by the Agricultural Development Society, Naini in early forties.

The Indian farmers gradually responded to farm mechanization technology especially after Green revolution in 1960s. High yielding varieties with assured irrigation and higher rate of application of fertilizer gave higher yields and better economic returns. This enabled the farmers to start adopting mechanization. The development of power thresher with integrated Bhusa making attachment and aspirator blower and mechanical sieves for grain and straw separation in 1960s was the major achievement of Indian engineers which was widely adopted by our farmers. Gradually demand for other farm machinery such as reaper and combine harvester also increased. Demand of tractors in the country was met through importation until 1961 when Eicher Tractors Ltd. and Tractors and

Farm Equipment Ltd started manufacturing tractors with foreign collaborations. To meet the additional demand , importation continued up to 1977. Meanwhile many other industries started manufacturing tractors with foreign know how such as Gujarat Tractors Ltd (1963), Escorts Ltd (1966), International Tractors (India) Ltd. (1966), and Hindustan Machine Tools Ltd (1977).

Punjab Tractors Ltd. started their production with indigenous technology in 1974. Many more industries started manufacturing tractors since then with indigenous and foreign know how.

FARM POWER IN AGRICULTURE

Unit farm power

During early sixties, a concept of farm power availability per hectare basis was used to indicate level of farm mechanization. The potential power availability was used as the measure and could not reflect critical constraint of farm power availability during peak requirements or the actual level of use. The total farm power availability from animate and mechanical sources in 1951-52 was 0.20 kW/ha which increased to 1 kW/ ha in 1996-97. Animate power contributed 60% of the total farm power in 1971-72 and mechanical and electrical together contributed only 40%. In 1996-97 the contribution from animate power reduced to 21% and from mechanical and electrical power it increased to 79%. The farm power input per unit cultivated land in India is still very low compared to South Korea of 7 kW/ha, Japan 14 kW/ha and United State of America of 6 kW/ha. It is evident that mechanical contributed about 78% to the total farm power but for tractive power it is only less than 30% and thus major use of mechanical power has been for stationary farm operations only. The growth in the ratio of availability of mechanical power to total power availability (per ha basis) and mechanical tractive power to total power availability in Indian agriculture indicates that although 78 per cent of farm power was available from mechanical power sources in 1996, only 29.6 per cent of mechanical tractive power offering higher level of technology was available to Indian farms.

Energy ratio for mechanization indicator

A major defect in considering power in quantifying mechanization is that it does not reflect the dimension of time and thus does not bring in the actual scenario in focus. Energy, from this point of view, is a better measure as it can define the actual quantity under use or, in other words, the potentiality of use of available power sources in real terms.

It may be seen that the total animate energy availability is decreasing over the years from 2937 MJ/ha in 1971 to 2505 MJ/ha in 1996- 97 due to decrease in availability of animal energy.

Human energy use increased from 1,331 to 1,525 MJ/ ha between 1996-97 and 1971. The animal energy use, however, decreased from 1,606 to 980 MJ/ha, indicating a negative annual growth of 1.9 per cent. Increased cost of maintenance of draught animal has been a main cause of this trend. Energy derived from electrical and diesel fuel used by mechanical power sources continued to increase with growth rate of 10.7 and 11.1 per cent respectively to supplement the additional energy need for increased land productivity. The total energy use for crop production increased from 3,282 to 8,773 MJ/ ha (annual growth rate of 3.6 per cent) during the period.

The share of mechanical energy over total energy used in crop production has increased from 11 per cent to 76 per cent during the same period. In order to meet the operational energy needs, the contribution of mechanical energy has to be increased substantially. To encourage use of tractors in agriculture, more diesel is required for the purpose. In order to achieve the production targets of food grains from 125 million hectare land, and assuming that the available energy is used for its production, the average specific energy availability on all India basis with present growth rate would be of the order of 5860 MJ/ tonne of food grain as compared to 5480 MJ/ tonne in 1996-97.

FARM POWER SOURCES IN AGRICULTURE

The use of farm machinery depends upon the farm power sources available in the country for various tractive and stationary operations. Human and animal power, the two 'renewable energy' or ' bio-energy' sources, have traditionally been used for various farm operations. The crops are protected from pests, diseases and from weeds, through the application of chemicals.

The application of these inputs is achieved through 'human power' in traditional agriculture. Animate power contributed 60.37% of the total farm power in 1971-72 and mechanical and electrical together contributed only 39.63%. In 2001- 02 the contribution from animate power has reduced to 16.38% and mechanical and electrical power, it increased to 83.62%.

This change in the relative contributions of the animate and nonanimate power sources has been due to the requirements of modern agriculture where timely application of crop production inputs is an important factor to maximize return on input investments. The nonanimate power equipment (like tractors, power tillers, self-propelled planting and harvesting machines, diesel engines and electric motors, etc) also helps farmers to increase cropping intensities through faster turnaround time between successive crops. In addition, the mechanical/Electric machines reduce the drudgery involved in the traditional methods of farming.

Agricultural Tractors

The sale of Agricultural tractors and other farm equipment has increased. To-day more than 250,000 tractors are manufactured every year by 13 manufacturers. These tractors are available in different horsepower ranges of less than 25 to more than 55 horse power. The share of various HP tractors in the recent years has been as follows.

Different sizes of tractors are manufactured in India ranging from less than 25 HP to more than 45 HP but most popular range is 31-35 HP.

The Tractor sales show that their demand is region specific. Punjab, Haryana and western UP constituted the major Tractor market in 1980's with 55-57% share of total All India sales. With increased Tractor population in these areas and good successive rains in 1990's, coupled with changes in cropping patterns like adoption of more profitable cash crops (Oilseeds, Pulses, etc.), and better prices, the sales in Gujarat, MP and Rajasthan have been seeing good growth. The contribution of these states improved from 20-22% in early 1990's to 30-32% by the close of the decade.

The share of eastern states, namely Bihar, Orissa, West Bengal and Assam had been consistently low at 3-5% due to various socio- economic, agro-climatic and other reasons. The credit availability to the farmers in this area has been another major reason for the slow growth in the eastern states. The tractor sales, since mid 1990's, have increased to about 10-12% of All India Sales.

FARM MACHINERY INDUSTRIES IN INDIA

The adoption of mechanization technology depends upon the local manufacture and after-sales-services besides credit and financial incentive provided by the Government. The manufacture of agricultural machinery in India is quite complex comprising from village artisans, tiny units, small scale industries to State AgroIndustrial Development Corporations and organized tractor, engine and processing equipment industries.

Traditional hand tools and bullock drawn implements are largely fabricated by village craftsmen and small scale industries. Organized sectors manufacture sophisticated machinery such as tractors, engines, mills and dairying equipment. The small-scale industries seldom have R&D facilities and they depend upon public institutions for technological support. They require not only drawings but also prototypes and technical guidance to manufacture the equipment. These industries however, upgrade the technology with experience.

Classification of Industries

The classification of industries in India is based on total capital investment (plant and machinery) rather number of workers employed. These are (i) village craftsmen, (ii) cottage industries, (iii) tiny industries, (iv) small scale industries, (v) medium scale industries and (vi) large scale industries. This classification was done to help the small-scale units through incentives and marketing support.

Village craftsmen

Village artisans are the main source of supply and repair and maintenance of hand tools and traditional implements are made by village craftsmen. These include implements and tools like khurpi, spade, sickle, local ploughs, bakhar, sowing devices, yokes, patela, leveller, oil ghanis, grinding wheels, hand mills, handoperated milk churning tools, winnowing devices, sieves, wooden storage structures, bullock carts, manual water lifting devices etc. If village artisans are properly trained they will accelerate the adoption of mechanization inputs due to their proximity with farmers.

Tiny and small-scale industries

The tiny and small scale units fabricate bulk of improved agricultural machinery such as ploughs, cultivators, disc ploughs and harrows, seed drills, planters, plant protection equipment, reaper harvesters, combine harvesters, threshers, cleaners, graders, mills, crushers, oil expellers, diesel engines, irrigation pumps, dairy machinery etc.

Agricultural machines are reserved for small-scale units. There are more than 18000 such units scattered all over the country but have concentration in selected regions.

Some of these units also fabricate implements and equipment for tractor and power tiller manufacturers. They may lack good machine tools and heat treatment facilities. Some of them are more organized and have better fabrication toolings nd thus are able to manufacture better quality machinery.

The bulk of the farm machinery is made by the small-scale industries. They use materials from mild steel to medium carbon steel. Heat treatment practices are generally inadequate except in few industrie manufacturing knife & tillage tools . Equipment manufactured by the SSI units includes Soil working tools, seeding & planting equipment, hand hoes, sprayers & dusters, harvesting & threshing equipment, like reapers, threshers , combines ,maize shellers , decorticators, cleaners, graders, mills, oil expellers etc.

Organized farm machinery industries

The medium scale and large scale industries manufacture diesel engines, electric motors, irrigation pumps ,sprayers and dusters , land development machinery, tractors, power tillers, post harvest and processing machinery and dairy equipment. There are 13 tractor, 2 power tiller, 200 diesel engine, 600 irrigation pump, 48 combine and 188 earthmoving machinery manufacturers . The marketing of agricultural machinery by these industries is through their network of dealerships and, therefore, these manufacturers are able to provide effective after-salesservice. These industries upgrade their product and process technologies through their own R&D efforts, in addition to technological support from external agencies.

Today, India is recognized as a leading country in the world for the development and manufacture of agricultural implements and equipment. The range of equipment includes, tractors, harvesting and threshing equipment, plant protection machines, irrigation and drainage pumps, sprinkler systems, land development machinery, dairy and agro-processing equipment, etc.

India is the exporting increasing volumes of these to various countries including USA, Africa, Asia, etc.

Standardization and quality

The Bureau of Indian Standards (BIS) with its network of centers and laboratories in the country is mandated to ensure quality manufacture and marketing of agricultural and industrial products in the country. The BIS formulates specifications of agricultural machinery and other appliances and prescribes test codes. The Bureau also issues ISI quality certificate marks to the products which meet technical specification as per BIS standards.

The Government of India has established Farm Machinery Training & Testing Centres for promotion of quality farm machinery. In case of agricultural machinery, requirement of quality certification is limited to the sale of agricultural machinery financed under Government Schemes. On certain items connected with safety and health hazards, it is mandatory to have minimum safety standards built into the design or in the installation of machinery during operation. Threshers, chaff cutters, sugarcane crushers etc. fall under this category. The agricultural machines manufactured by the organized sector like tractors ,earth moving machinery, irrigation equipment, plant protection, dairy equipment, processing machinery etc. are certified for their quality by BIS.

Testing and evaluation

The adoption of agricultural machinery is greatly influenced by the quality and after sales available to the farmers. Since manufacture of agricultural machinery is reserved for small-scale industries, the quality is affected by the manufacturing technology adopted by them. Testing and evaluation helps in up-gradation and quality production of machinery. R&D institutions and quality certification agencies conduct the T&E. Testing and evaluation is conducted on newly developed equipment and during its serial production, to facilitate and ensure quality, reliability, durability, functional ease, comfort in operation and cost of operation. Testing is conducted with well defined standard parameters, defined in BIS, ISO, or OECD standards, and where as evaluation is done to measure the performance under simulated or field conditions for the parameters for which the equipment has been designed.

FUTURE THRUSTS IN AGRICULTURAL MECHANIZATION

India is a large country with wide-agro ecological diversity having predominance of rainfed agriculture, with irrigated agriculture limited to 34% only. Farm holdings are small due to higher population density and land fragmentation will continue due to 'Laws of Inheritance' and 'Hindu Succession Act'. Majority of the farmers have limited surplus money to modernize farms or to invest in improved inputs. Draught animals and increasing agricultural workers population may remain to be the major source of farm power for soil manipulation and for crop handling, particularly in Hill and Mountain regions. Mechanical power for tillage, irrigation, harvesting and threshing will be preferred, including on custom hiring basis. As a result of GATT agreement, prospects of agro-export are likely to increase and product quality standards stipulated under WTO would encourage more and more farmers to adopt modern agricultural production technologies. The future agricultural mechanization technology package therefore may have to;

- be eco-friendly utilizing land water and bioresource catering to the varied group of farm holders,
- facilitate farming operations which are arduous and hazardous,
- increase productivity and conserve resources through effective utilization of chemical, biological and mechanical inputs, and
- modernize commercial agriculture to facilitate agro-export.

Keeping above objectives, the mechanization policy may have to be distinctly different to serve hill agriculture, low lying water logged soils, rainfed and irrigated lands and regions having agro export potential.

FUTURE PERSPECTIVE IN AGRICULTURAL MACHINERY MANUFACTURE

Equipment for tillage, sowing, irrigation, plant protection and threshing have been widely accepted by the farmers in India. Draught animal and human power in India will continue to be used, but these are inadequate to ensure timeliness of agricultural operations. Even farmers with small holdings utilize selected improved farm equipment, including through custom hiring. The future mechanization strategy may have to be based on agro-ecological diversity and economic disparity of the farmers. The present trend in agricultural mechanization is for high capacity machines to be used on custom hiring and for contractual field operations.

Rice mechanization, sugarcane mechanization, cotton mechanization, potato mechanization, horticulture mechanization, green house and covered cultivation, drip and micro irrigation are new emerging areas which need attention of Agricultural Engineering Institutions and industries for their development, production and marketing.

Water is a scarce commodity and in future with increasing demand for more irrigation water, concerted efforts will be needed for controlled application of water through drip, sprinkler and micro-sprinkler systems to economize use of water and improving water use efficiency With the shift in agriculture towards diversification and agri-business, substantial areas will go under horticultural crops. This will also help to export good quality high value agri-products for better returns to farmers and to earn more foreign exchange. The green house technology offers ample scope for increasing productivity particularly of high value cash crops like exotic fruits, flowers and bio-tech plants.

Presently little effort has been made to mechanize hill agriculture, where there is tremendous potential of growing horticultural crops, flowers etc. In future this calls for developing appropriate technologies for mechanization.

In order to enforce quality, reliability and safety in the ma n u fact ure of agricultural implements , manufacturing of critical components need to be standardized and encouraged for mass production by medium and large scale manufacturers. Keeping long s t anding demand of farmers and the Mini stry of Agriculture and on the recommendation of the Advisory Committee of the Ministry of Industries, the Union Budget of India 1998-99, announced the exclusion of farm implements and tools from the list of items reserved for manufacture by small scale industries sector to enable the farmers to get benefit of wider range of implements and tools at competitive prices, and with requisite aftersale-service. The decision of the Government of India to de-reserve the manufacture of farm machinery will help the organized sector to bring latest farm machinery technology for accelerated adoption by the farmers. The small-scale industries in turn will adopt the technology for local manufacturing at a much lower cost. This will help the small-scale sector to become more competitive and to enlarge their market size.

A RETURN OF THE THRESHING RING?

A Case Study of Machinery and Labor-Sharing in Midwestern Farms

Machinery-sharing provides an alternative for smaller producers to obtain the efficiencies of large farming operations and remain competitive in an increasingly concentrated agricultural industry. This research uses a multiple case study design to examine the motivations for sharing equipment and labor among farms and to better understand how group members handle the transaction costs of sharing. Our case evidence finds that in addition to cost savings, access to reliable labor is an important motivation for participating in a sharing arrangement. Trust and frequent communication among group members helps to minimize the transaction costs incurred from sharing.

A number of market forces favoring larger farm operations are driving U.S. agricultural production toward a larger scale. Among them is escalating input costs, especially machinery costs. Estimated per-farm costs for tractors and other self-propelled machinery in Midwestern states rose 54% in real terms over the past decade, from roughly $4400 per farm in 1998 to $6800 per farm by 2008 (NASS, 2009). Because machinery is a "lumpy input" that must be adopted in discrete amounts, expanding production to take advantage of size economies is one way to overcome the associated rise in fixed costs (Johnson and Ruttan, 1994).

Although many producers have increased their farm size, this strategy is not always appropriate or feasible. Insufficient access to land and capital can limit expansion. Expanding production may increase risk as a result of larger investments in a small set of related commodities (Roe, 2005). Also, there are limits to farm size growth. The weather-dependent nature of farming and the high level of uncertainty involved in farming requires quick decision-making by farm managers, a role that cannot typically be delegated to workers (Johnson and Ruttan, 1994). The prevalence of family farms has been explained as a result of their ability to minimize transaction costs, particularly the costs of monitoring workers.

An alternative strategy to deal with escalating farm machinery and equipment costs is to share machinery.1 Many farmers occasionally share a piece of equipment or trade a few days labor and it is common for neighbors to help when a farmer is sick or injured. However, there is evidence that a number of U.S. farmers are revisiting the practice of sharing equipment and labor with other producers on a more routine basis. For example, the original idea behind the agricultural leasing firm MachineryLink was to share machinery between farmers over a geographic distance to take advantage of differences in growing seasons across regions.

Machinery-sharing rings and other farm-level cooperative arrangements are more common in Europe and Canada.2 Studies of these machinery cooperatives have documented machinery costs savings for members as well as several related benefits. These include access to specialized or more efficient machinery; scale economies; an ability to draw on the experience, labor, and ideas of other members; the possibility of labor or task specialization; access to volume discounts on other inputs; risk-sharing; and environmental benefits from reduced input use. Potential drawbacks identified include a loss of timeliness in field operations, decreased autonomy in decisionmaking, more complex management, potential problems with lenders and split lines of credit, and difficulties in unwinding the arrangement.

The research on resource-sharing arrangements in the U.S. is sparse. A few University Extension guides address machinery-sharing as one of several options for controlling equipment costs on the farm (for example, see Edwards, 2001; Weness, 2001). Lawless, Cropp, and Harris (1996) address potential advantages and disadvantages of various legal business structures for multifamily dairy operations in Wisconsin but do not analyze other aspects of these arrangements. Stofferhn's (2004) survey of North Dakota farmers and ranchers found roughly 11% of respondents indicating a willingness to share equipment or labor with a neighbor but did not document examples of the actual practice. Evidence of cooperative agreements between producers for sharing equipment is therefore largely anecdotal and not well understood.

This research seeks to 1) describe the motivations of farmers who have entered into agreements to share machinery and/or labor with one another; and 2) document the ways in which these groups have experienced and addressed the transaction costs (timeliness, group decision-making, moral hazard problems) of sharing. Given the limited amount of information on machinery-sharing groups in the U.S., this research is exploratory in nature. We use a case study approach that complements existing theoretical and empirical research by providing more detailed information about how existing machinery-sharing arrangements function.

The article proceeds as follows. We outline a conceptual framework of the benefits and costs of sharing equipment and labor between farms in the next section. We discuss the data collection methods next. Presentation and discussion of the internal and external economies of scale and transaction costs uncovered in the case studies follow. The final section concludes with a summary of research findings.

A Conceptual Framework for Machinery-Sharing

Participation in a machinery-sharing arrangement involves a tradeoff between access to internal and external economies of scale and increased transaction costs (Allen and Lueck, 1998; Valentinov, 2007). Internal economies of scale arise from improvements in technological efficiency; that is an ability to produce more output with the same inputs or to produce the same output with fewer inputs (Varian, 1992). Working in a group to share machinery tends to increase the number of acres serviced by the machinery, reducing inputs, and average costs for a given amount of output. Sharing can therefore make newer, larger, more technologically advanced equipment economical. In addition, group members can improve labor productivity by coordinating tasks to reduce duplication and allow for task specialization.

External economies of scale are related to advantages larger farms may have in accessing inputs; obtaining and negotiating terms of credit, storage, services; and marketing and distribution opportunities (Johnson and Ruttan, 1994). For example, larger farms can negotiate volume discounts on inputs (McBride, 2003). Larger farms might be able to attract specialty contracts that pay premiums for delivery of greater quantities of product. Like marketing cooperatives that obtain higher retail prices through quality assurance, smaller farmer groups may be able to successfully coordinate production practices such as planting and harvest times to maximize quality specifications (Sexton and Iskow, 1988).

Transaction costs related to farming in a partnership or group include timeliness costs, monitoring costs, and costs related to group decision-making. Timeliness costs may occur when a field operation cannot be performed at the optimal time (de Toro and Hansson, 2004; Lars?n, 2007). Monitoring costs may arise to ensure that other members are careful with the shared equipment (an asset moral hazard problem) or are contributing an agreed-on number of hours of work when labor is shared (an effort moral hazard problem) (Allen and Lueck, 1998; Lars?n, 2007). In addition, collective decision making may be costly, particularly when the interests of group members are not well-aligned (Hansmann, 1996).

Allen and Lueck (1998) model the optimal farm organization (family farm, partnership, or corporation) as a tradeoff between gains from task specialization and lower capital costs and increased transaction costs, particularly moral hazard costs. They empirically test their theory by examining the prevalence of organizational forms across production types (different crops, different regions, and over time). Their analysis shows that when production is seasonal and there are many stages to production with few tasks, the gains from specialization are limited while monitoring costs are high. In this type of production such as with grain farming, family farms are the dominant organizational form and partnerships are relatively rare.

Larsn (2007) builds on Allen and Lueck's model to incorporate social norms, peer pressure, and dynamics (repeated interaction) as factors that might mitigate the moral hazard costs involved in partnerships. Her survey of 640 Swedish farms found 62% participating in some form of partnership arrangement. The partnership farms in this study were characterized by a high degree of trust and Larsn concludes that the perceived moral hazard problems were negligible.

It is important to note that both the benefits and the costs of sharing relative to farming alone will vary with the size and scope of the sharing and the parties involved. Sharing only a combine with a neighbor during harvest may lower machinery costs while requiring some amount of coordination between partners but is unlikely to result in volume discounts on other inputs or increased output prices. In contrast, a group of many producers farming in a more integrated manner year-round might encounter both significantly greater benefits as well as significantly greater coordination costs.

STUDY DESIGN AND DATA

We examine the tradeoffs between internal and external economies of scale and transaction costs in a set of 10 cases of machinery-sharing groups primarily located in the Midwestern U.S. We focused our study on farms with similar production types, primarily grain production (corn, soybeans, wheat), but chose cases to represent variety in the degree of inter-farm cooperation to understand how changes in the scope of sharing affect transaction costs. Additionally, we sought cases in which some, if not all, members were unrelated, believing that kinship ties might confound the analysis of group dynamics.

Case study approaches are well-suited for gaining an in-depth understanding of current events and for asking the types of "why" and "how" questions at the heart of this study (Kennedy and Luzar, 1999; Sterns, Schweikhardt and Peterson, 1998; Westgren and Zering, 1998; Yin, 2003). Analysis of organizational forms in secondary data like the Agricultural Resource Management Survey is hampered by the fact that most machinery-sharing arrangements are informal and therefore not reflected in the data. Given the relatively small number of machinery sharing groups in the U.S., data collection through a random survey of producers might not identify many, if any, groups for analysis.

The case study approach does have its limitations. We cannot calculate an incidence of machinery-sharing, for example. In addition, our findings do not generalize to the population, but rather to a set of theoretical propositions (Yin, 2003). The case approach taken in this study does allow us to illustrate the range of organizational forms and strategies used to share machinery in Midwestern grain farms, to document differences across cases, and to describe any systematic patterns in these differences.

We used a multiple case design in which the unit of analysis was the machinery-sharing group. Potential case study subjects were identified through a web-based survey of University Extension professionals in five Midwestern states. Respondents were asked to identify local groups that fit the following description: "independent farms that are participating in cooperative arrangements to share resources for production, yet retaining decision making sovereignty over their assets and labor." University Extension staff responding to the survey identified 50 groups in five states (Iowa, Illinois, Wisconsin, Nebraska, and Indiana) sharing equipment and/ or labor.

The majority of groups identified in the survey had only two or three members (39 of 50). Half of the groups contained no family members, whereas 30% of the groups were comprised solely of family members and 18% combined family and nonfamily members. Slightly more than half of the groups had existed for more than 5 years. Although the groups were engaged predominantly in grain production, cooperative arrangements among vegetable, livestock, and fruit growers were also identified.

The web survey discovered a wide range of cooperative arrangements. Seventy percent of groups identified share both machinery and labor. Fourteen percent were reported to share only machinery, whereas 10% reportedly share only labor. A variety of equipment types were identified in the respondents' comments. A number of groups share sprayers, combines, and harvest equipment. A few groups share equipment for haying, manure hauling, seeding, and irrigation. Some producers only share the cost of a single piece of equipment. Other groups share whole machinery sets (e.g., combine, tractor, planter, and sprayer) as well as labor for operating the equipment. Some others share not only equipment and labor, but also purchase inputs and market output as a group.

The formality of the sharing arrangements varied from simple verbal agreements to written contracts and formal business structures. Over half of the groups identified have a verbal agreement for sharing resources. Only four of the groups identified were known to have a written agreement. However, for a significant portion of the groups (40%), the type of agreement was unknown by Extension staff at the time of the survey.

Based on the web survey results, we developed the two-dimensional taxonomy of cooperative arrangements shown. The vertical dimension represents the degree of cooperation, ranging from sharing only one piece of machinery to a highly integrated case in which producers share all equipment, purchase inputs jointly, and market at least some output together. The horizontal dimension represents the degree of formality of the agreement ranging from a verbal agreement to a formal business structure.

Cases focused on examples of sharing in Midwestern grain operations, but we attempted to identify cases representing the spectrum of formality in arrangements and degree of cooperation. Selecting cases to represent this diversity was meant to facilitate comparisons among groups. Ten case study analyses based on in-person interviews of producer groups who share or have shared resources among farms were completed between June 2004 and January 2005 It places these cases in the context of our taxonomy. Potential participants were contacted by telephone and by mail to solicit their participation in the study. With the exception of LMC whose members were interviewed during a visit to our university, we conducted interviews with group members "on-site" in a neutral location such as a local Extension office or hotel meeting room in their local community. Once a meeting time and location were agreed on, a pre-interview questionnaire was sent to the group contact requesting basic information about group history and members.

We conducted group interviews with as many group members as possible using interview guides, basic outlines of the general topics, and open-ended questions to be discussed.6 Interviews lasted approximately 2 hours and each participant received $50 to compensate them for their time. The interviews were taperecorded when possible and later transcribed. Investigators took notes during the interview and wrote summaries of the conversations as soon as possible after the interviews occurred. As a result of a concern that members would not feel comfortable expressing negative opinions about their arrangement in the presence of their partners, we provided each member with an exit questionnaire to be completed at a later time and mailed to the investigators in a stamped, addressed envelope we provided. The exit questionnaire solicited members' opinions about the most negative aspects of their equipment-sharing practices and requested any other information they felt relevant to the study.7 The notes of the investigators, transcripts, and completed questionnaires served as the basis for the analysis of the case study findings.

The degree of labor-sharing varied from none, in the case of a long-distance equipment-sharing arrangement, to a few weeks during harvest season to fully integrated labor operations year-round. Agreements between group members were of two basic types. Groups involving fewer than four members, regardless of the number of equipment pieces or amount of labor shared, predominantly had only a verbal agreement. Groups with a significant scale and scope of operations typically had a written contract, and most had formed a business entity (e.g., LLC). Nine of the groups were operating at the time of the interviews; one group had dissolved.

INTERNAL ECONOMIES OF SCALE

In nearly all cases, sharing helped groups improve technical efficiency. Sharing machinery and working together reduced per-acre equipment and labor costs in the majority of cases. In others, machine costs per acre remained comparable to farming individually but allowed members access to newer, technologically advanced machinery. Producers reported an increased speed of operations as a result of higher-capacity machines, especially during harvest, fewer breakdowns, larger pools of labor, and more efficient use of labor. As one producer put it, "Three go twice as fast as two." Some producers stated that, as a result of the cost and time savings of their cooperative agreement, they were able to significantly expand the acreage of their crop operation or expand the size of their livestock operation.

Somewhat surprisingly, labor, more so than machinery costs, motivated sharing in a number of cases. Many producers rely on family members or retired neighbors to help during busy times, but finding reliable, skilled, and seasonal labor can be a challenge. I0 Working with nearby producers offers one solution to this problem. Several groups reported that sharing eliminated their need to hire outside labor and improved efficiency of operations, particularly during harvest. Operating as a group allows for some specialization. To the extent that group members can take advantage of their complementarities, sharing can reduce the amount of individual effort required. In several cases, producers reported assuming responsibility for tasks they enjoyed or for which they had special training or knowledge. For example, in the case of Zimmerman and Erickson, Zimmerman takes care of most machinery maintenance and repairs because he is a mechanic by training. Erickson, who has a commercial driver's license, handles much of the hauling and scheduling with their local cooperative.

The exception was the case of Bennett, Taylor, and Nelson. Rather than improving efficiency through group efforts, in this case sharing seemed to reduce it. This failed cooperative effort was stymied by the group's inability to agree on farming practices and scheduling (specifically how, when, and by whom field work would be done) as well as the purchase of an undersized combine for their combined number of acres.

EXTERNAL ECONOMICS OF SCALE

The main external economies of scale reported by groups related to improved negotiating power. Several groups reported advantages with regard to renting land. Panhandle Farms felt their ability to plant and harvest in a timely fashion provided an edge in obtaining crop share leases, which are desirable because they require less capital and carry less financial risk than cash rent leases. The Spaulding family noted an advantage from their group efforts with regard to landlords. They jointly own equipment for tile work, which allows them to make improvements to their rented farms more inexpensively than the going rate.

Another perceived advantage of group size appears to be special treatment from machinery dealers, input suppliers, and local elevators. When the members of Progressive Farmers encountered problems with a new combine, the manufacturer sent two engineers directly to their farms to fix it. They felt this was a direct consequence of their size. The Spauldings reported that the local elevator occasionally extended its hours to accommodate additional deliveries from the group. Anderson and Parker also felt their larger-volume purchases resulted in improved service from their dealers.

In contrast, the cases provided very little evidence of marketing opportunities being exploited. Only Panhandle Farms and LMC jointly marketed any of their production. Only LMC had attempted to coordinate production practices to attract price premiums. LMC expanded its operations into a variety of subsidiaries, including a seed-cleaning business and a specialty crops export venture. Parker and Anderson did suggest that a major benefit of their partnership was having someone to share ideas with: "Two heads are better than one." They felt the opportunity to routinely discuss marketing strategies improved their bottom line, although they continue to market their crop separately.

Even joint input buying was problematic in many cases because members maintained loyalties to different seed and chemical dealers. When groups did report coordinated buying of inputs, they generally claimed savings. For example, the Spauldings reported that coordinated purchases resulted in a discount of $3 per acre on chemicals and several free bags of seed each year. Anderson and Parker estimated a 15-20% savings on seed and chemicals from joint purchases.

Despite the lack of coordination observed in input buying and marketing among these groups, if opportunities arose, these groups would be well positioned to take advantage of them. Their history of successfully coordinating group machinery use gives them experience managing group dynamics other producers frequently lack. As in the case of LMC, and to a lesser extent Panhandle Farms and Zimmerman and Erickson, such improved coordination may evolve over time as the group gains experience working together.

Transaction Costs

Sharing equipment and labor involves transactions costs. Members must coordinate schedules, production practices, and, in some cases, even seed varieties. They must make joint decisions about what type of equipment to share, when to trade, how to operate as a group, how to handle repairs and regular maintenance, and a variety of other issues. The more integrated the operations of a group, the more likely additional recordkeeping is required. These costs seem to be higher in the beginning stages of the sharing arrangement. Once groups have operated for some time, many decisions become routine.

Timeliness Costs. Several groups reported that they viewed their separate land holdings as one operation for purpose of deciding which fields to work and when. Groups tried to strike a balance between optimal timing and fairness to individual members. One member of the Spauldings explained their method for making decisions about timing in this way: "Whatever's ready first. I don't think that's ever been a question. Sometimes we go around and do a little bit of everybody's. We don't do all of one person's first." Progressive Farmers uses a more systematic approach. Although their land is not adjacent, it has a general north-south orientation. Initially, the group began planting corn in the north working southward. They would then retrace their path from south to north when planting soybeans. Although this approach seemed efficient, they felt it gave the farmland in the middle an unfair advantage. Now the group uses a circular planting scheme and rotates the starting point each year. During harvest, they combine fields as they are ready while making sure they harvest some of each member's land as they go, not leaving any one member's land for last.

In some cases, differences in land types or locations facilitate sharing. The heart of the longdistance arrangement between the Fergusons and Duncans is their ability to exploit variation in growing seasons between their farms. Their shared combine begins the season on Ferguson's North Dakota farm for small grain harvest. The machine moves to the Duncans' Minnesota farm for corn and soybean harvest on September 20, as specified in their operating agreement. This September 20 date is flexible depending on the harvest conditions in any given year. They try to be flexible and communicate with each other to work out solutions that benefit both parties.

Even neighboring farms can take advantage of differences in their land to mitigate timeliness issues. In Zimmerman and Erickson's partnership, Erickson's property is low and consists of river and creek bottom land, whereas Zimmerman has mostly hill ground. Therefore, Zimmerman's property can usually be worked sooner than Erickson's. They realized that by working together they could avoid some of the timing problems associated with working bottom ground before it was ready and working the upland more slowly than would be desirable. In other words, if the fields were planted as if each producer's land were part of a single, larger operation, the outcome for both producers would be better.

Finally, several groups stated that taking account of one another's decisions about hybrids helped reduced timeliness issues, even if they did not make seed decisions jointly. One producer described it this way: "He doesn't tell me what specific hybrids and I don't tell him what specific hybrids. We know that in general we need to have somewhere in the neighborhood of 1 10 day hybrid on this end of the stick and somewhere in the general neighborhood of 1 12 days on this end of the stick."

Monitoring Costs (Effort and Carefulness). In the majority of the cases, a sense of trust and shared values among partners rendered the effort moral hazard problem moot. Most groups interviewed did not track hours contributed to the group effort. In many cases, they worked together on a daily basis and could therefore "monitor" each other's efforts. Overall, statements such as "You don't ask somebody to do something that you wouldn't do yourself" reflected the sense that all partners were committed to the agreement and worked hard to make it successful.

Two notable exceptions to this finding are provided by the cases of the Spauldings and Bennett, Taylor, and Nelson. At the time of the interview, members of the Spauldings group recognized they would need to devise a system to account for members' unequal contributions of hours. Specifically, one of the younger members ran a growing agriculture-related business in addition to farming. Because the busy time for both operations coincides, contributing his share of hours to the farming operation in the spring and fall was becoming increasingly difficult for him. "He's getting big enough now, you are going to have to make some decisions." Another member, a son-in-law who worked fulltime off-farm and was an inexperienced farm equipment operator, paid $20-25 an acre rent to help compensate for his reduced labor contribution.

In addition to sharing equipment, an important part of the Bennett, Taylor, and Nelson's agreement was shared labor. Going into the arrangement, they thought they shared a similar work ethic and view toward farming. As a result, they did not track hours and did not compensate one another for their time. They encountered problems when Bennett hired his brother to complete fall tillage on his land while he worked with the others harvesting. Because of his fulltime off-farm job, Bennett had limited time off during harvest. Hiring his brother was the only way he felt he could complete both his harvest and fall tillage before he needed to return to his job in town. After harvest, when Nelson and Taylor still needed to complete their fall tillage, the fact that Bennett was not available to help created resentment.

23

PRECISION AGRICULTURE

Precision agriculture (PA) is an approach to farm management that uses information technology (IT) to ensure that the crops and soil receive exactly what they need for optimum health and productivity. The goal of PA is to ensure profitability, sustainability and protection of the environment. PA is also known as satellite agriculture, as-needed farming and site-specific crop management (SSCM).

Precision agriculture relies upon specialized equipment, software and IT services. The approach includes accessing real-time data about the conditions of the crops, soil and ambient air, along with other relevant information such as hyper-local weather predictions, labor costs and equipment availability. Predictive analyticssoftware uses the data to provide farmers with guidance about crop rotation, optimal planting times, harvesting times and soil management.

Sensors in fields measure the moisture content and temperature of the soil and surrounding air. Satellites and robotic drones provide farmers with real-time images of individual plants. Information from those images can be processed and integrated with sensor and other data to yield guidance for immediate and future decisions, such as precisely what fields to water and when or where to plant a particular crop.

Agricultural control centers integrate sensor dataand imaging input with other data, providing farmers with the ability to identify fields that require treatment and determine the optimum amount of water, fertilizers and pesticides to apply. This helps the farmer avoid wasting resources and prevent run-off, ensuring that the soil has just the right amount of additives for optimum health, while also reducing costs and controlling the farm's environmental impact.

In the past, precision agriculture was limited to larger operations which could support the IT infrastructure and other technology resources required to fully implement and benefit from the benefits of precision agriculture. Today, however, mobile apps, smart sensors, drones and cloud computing makes precision agriculture possible for farming cooperatives and even small family farms. Precision agriculture was born with the introduction of GPS guidance for tractors in the early 1990s, and the adoption of this technology is now so widespread globally that it's probably the most commonly-used example of precision ag today. John Deere was the first to introduce this technology using GPS location data from satellites. A GPS-connected controller in a farmer's tractor automatically steers the equipment based on the coordinates of a field. This reduces steering errors by drivers and therefore any overlap passes on the field. In turn, this results in less wasted seed, fertilizer, fuel, and time.

HIGHLIGHTS OF PRECISION AGRICULTURE

Precision agriculture, also called "farming by the foot", can be defined as a comprehensive system designed to optimize agricultural production through the application of crop information, advanced technology and management practices. A truly comprehensive approach to precision agriculture begins with crop planning and includes tillage, planting, chemical applications, harvesting, and post harvest processing of the crop. Basically precision agriculture is using advanced technology in culmination with existing agriculture equipment to maximize yield while minimizing losses. These losses come in many forms; some include overspending on fertilizer, seed, pesticides, and herbicides due to poor placement. This can be minimized by precisely distributing them because they tend to be dispersed with overlaps occurring or skips taking place. With the use of precision ag the farmer also maximizes his efficiency and makes the most of his time spent in the field.

The rudimentary feature of precision ag is the Global Positioning System (GPS). GPS operates using several satellites strategically placed in different orbits in space. These satellites transmit signals to a receiver on earth and, by a mathematical expression known as trilateration, let the receiver know exactly where it is. In precision agriculture, the receiver is the round object usually located on top of the tractor cab. The receiver communicates with the computer system located in the tractor cab and the operator programs the computer to carry out a task specific to the type of farming he/she is engaged in. For example, the farmer can tell the tractor to drive a straight path across the field and push a button and the tractor will drive itself along the course until the farmer is ready to take over again or turn the tractor around. The best GPS systems today that are integrated into farming have accuracy within a 3 cm range. It is pretty phenomenal knowing exactly where you are within 3 cm. across a 100 acre field.

Some of the more intricate aspects of precision agriculture include site specific crop management and variable rate technology. Site specific crop management uses a yield map of the field to predict and prescribe fertilization and pesticide application to different parts of the field as needed. This has been made possible by recent innovation in information and technology such as microcomputers, geographic information systems, positioning technologies (GPS), and automatic control of farm machinery. Using this technology one can avoid having to put a uniform distribution of fertilizer pesticides, or seeds across a field and wasting products in areas that may not necessarily need it. This is where the slogan "farming by the foot" was derived because the yield maps can provide very small scale application needs almost down to the square foot.

Fig 23.1

A yield map can be generated using GPS, a computer program such as ArcGIS, and a load cell or impact plate on the implement all working collectively. The impact plate or load cell will measure the mass of a crop being harvested in the field at the same time the GPS system is recording exactly where the tractor is in the field. The yield at this specific location is associated with a GPS position and saved in the computer. It then can be uploaded into ArcGIS and a map made from the information indicating the amount of grain harvested at each point in the field.

Precision agriculture encapsulates uses of GPS systems, farming equipment, and computer systems and makes it almost a must for a farmer in today's society who farms even a moderate amount of acreage. The benefits of precision ag can save a farmer 8% to 30% a year which can translate to about $30,000 or more. At this rate, a farmer can equip all of his implements with this technology and it will pay for itself in a few short years. Clearly, a modern day farmer cannot afford to be without precision agriculture equipment.

Agriculture dominates the world's land use decisions. The urgent need for doubling farm production over the next 25 years on less land with less water through further intensification would inevitably involve substantial social, economic, and environmental costs. Identification of tools to minimize such costs through enhanced productivity and economic profits while simultaneously conserving the environment is, therefore, crucial. Precision agriculture (PA) is one of such tools catching worldwide attention since the early 1990s.

Precision agriculture (PA), satellite farming or site specific crop management (SSCM) is a farming management concept based on observing, measuring and responding to inter and intra-field variability in crops. The goal of precision agriculture research is to define a decision support system(DSS) for whole farm management with the goal of optimizing returns on inputs while preserving resources.

Among these many approaches is a phytogeomorphological approach which ties multi-year crop growth stability/characteristics to topological terrain attributes. The interest in the phytogeomorphological approach stems from the fact that the geomorphology component typically dictates the hydrology of the farm field.

The practice of precision agriculture has been enabled by the advent of GPS and GNSS. The farmer's and/or researcher's ability to locate their precise position in a field allows for the creation of maps of the spatial variability of as many variables as can be measured (e.g. crop yield, terrain features/topography, organic matter content, moisture levels, nitrogen levels, pH, EC, Mg, K, and others). Similar data is collected by sensor arrays mounted on GPS-equipped combine harvesters. These arrays consist of real-time sensors that measure everything from chlorophyll levels to plant water status, along with multispectral imagery. This data is used in conjunction with satellite imagery by variable rate technology (VRT) including seeders, sprayers, etc. to optimally distribute resources.

Precision agriculture has also been enabled by unmanned aerial vehicles like the DJI Phantom which are relatively inexpensive and can be operated by novice pilots. These agricultural drones can be equipped with hyper spectral or RGB cameras to capture many images of a field that can be processed using photo gram metric methods to create orthophotos and NDVI maps. These drones are capable of capturing several metric points of the land that can later be used to deliver proper water and fertilization to crops.

PRECISION AGRONOMICS

Precision agronomics is another important term related to the combining of methodology with technology. At its core, it's about providing more accurate farming techniques for planting and growing crops. Precision agronomics can involve any of the following elements:

Variable rate technology (VRT) – VRT refers to any technology that enables the variable application of inputs and allows farmers to control the amount of inputs they apply in a specific location. The basic components of this technology include a computer, software, a controller and a differential global positioning system (DGPS). There are three basic approaches to using VRT – map-based, sensor-based and manual. The adoption of variable rate technology is currently estimated at 15% in North America and is expected to continue to grow rapidly over the next five years.

GPS soil sampling – Testing a field's soil reveals available nutrients, pH level, and a range of other data that is important for making informed and profitable decisions. In essence, soil sampling allows growers to consider productivity differences within a field and formulate a plan that takes these differences into account. Collection and sampling services that are worth the effort will allow the data to be used for input for variable rate applications for optimizing seeding and fertilizer.

Computer-based applications – Computer applications can be used to create precise farm plans, field maps, crop scouting and yield maps. This, in turn, allows for the more precise application of inputs such as pesticides, herbicides, and fertilizers, thus helping to reduce expenses, produce higher yields and create a more environmentally-friendly operation. The challenge with these software systems is they sometimes deliver a narrow value that doesn't allow data to be used for making bigger farm decisions, especially with the support of an expert. Another concern with many software applications is poor user interfaces, and the inability to integrate the information they provide with other data sources to enrich and show significant value to farmers.

Remote sensing technology – Remote sensing technology has been in use in agriculture since the late 1960s. It can be an invaluable tool when it comes to monitoring and managing land, water, and other resources. It can help determine everything from what factors may be stressing a crop at a specific point in time to estimating the amount of moisture in the soil. This data enriches decision-making on the farm and can come from several sources including drones and satellites.

At its most basic level, precision agronomics takes the role of an agronomist and helps make the methods they use more accurate and scalable.

The primary aim of precision agriculture and precision agronomics is to ensure profitability, efficiency, and sustainability while protecting the environment. This is achieved by using the big data gathered by this technology to guide both immediate and future decisions on everything from where in the field to apply a particular rate, to when it's best to apply chemical, fertilizer or seed.

While precision agriculture principles have been around for more than 25 years, it's only been over the past decade that they have become mainstream due to technological advancements and the adoption of other, broader technologies. The adoption of mobile devices, access to high-speed internet, low cost and reliable satellites – for positioning and imagery — and farm equipment that's optimized for precision agriculture by the manufacturer, are some of the key technologies characterizing the trend for precision agriculture. Some experts have suggested that more than 50% of today's farmers use at least one precision farming practice.

HISTORY

Precision agriculture is a key component of the third wave of modern agricultural revolutions. The first agricultural revolution was the increase of mechanized agriculture, from 1900 to 1930. Each farmer produced enough food to feed about 26 people during this time. The 1990s prompted the Green Revolution with new methods of genetic modification, which led to each farmer feeding about 155 people. It is expected that by 2050, the global population will reach about 9.6 billion, and food production must effectively double from current levels in order to feed every mouth. With new technological advancements in the agricultural revolution of precision farming, each farmer will be able to feed 265 people on the same acreage.

The first wave of the precision agricultural revolution will come in the forms of satellite and aerial imagery, weather prediction, variable rate fertilizer application, and crop health indicators. The second wave will aggregate the machine data for even more precise planting, topographical mapping, and soil data.

Precision agriculture aims to optimize field-level management with regard to:

- crop science: by matching farming practices more closely to crop needs (e.g. fertilizer inputs);
- environmental protection: by reducing environmental risks and footprint of farming (e.g. limiting leaching of nitrogen);
- economics: by boosting competitiveness through more efficient practices (e.g. improved management of fertilizer usage and other inputs).

Precision agriculture also provides farmers with a wealth of information to:

- build up a record of their farm
- improve decision-making
- foster greater traceability
- enhance marketing of farm products
- improve lease arrangements and relationship with landlords
- enhance the inherent quality of farm products (e.g. protein level in bread-flour wheat)

Prescriptive planting

Prescriptive planting is a type of farming system that delivers data-driven planting advice that can determine variable planting rates to accommodate varying conditions across a single field, in order to maximize yield. It has been described as "Big Data on the farm." Monsanto, DuPont and others are launching this technology in the US.

Tools

Precision agriculture is usually done as a four-stage process to observe spatial variability:

Data collection

Geolocating a field enables the farmer to overlay information gathered from analysis of soils and residual nitrogen, and information on previous crops and soil resistivity. Relocation is done in two ways:

- The field is delineated using an in-vehicle GPS receiver as the farmer drives a tractor around the field.
- The field is delineated on a basemap derived from aerial or satellite imagery. The base images must have the right level of resolution and geometric quality to ensure that geolocation is sufficiently accurate.

Variables

Intra and inter-field variability may result from a number of factors. These include climatic conditions (hail, drought, rain, etc.), soils (texture, depth, nitrogen levels), cropping practices (no-till farming), weeds and disease. Permanent indicators—chiefly soil indicators—provide farmers with information about the main environmental constants. Point indicators allow them to track a crop's status, i.e., to see whether diseases are developing, if the crop is suffering from water stress, nitrogen stress, or lodging, whether it has been damaged by ice and so on. This information may come from weather stations and other sensors (soil electrical resistivity, detection with the naked eye, satellite imagery, etc.). Soil resistivity measurements combined with soil analysis make it possible to measure moisture content. Soil resistivity is also a relatively simple and cheap measurement.

Soil apparent Electrical Conductivity (ECa)is another chief parameter that can provide a measure of the spatial differences associated with soil physical and chemical properties, which for paddy soil may be a measure of soil suitability for crop growth, its water demand and its productivity...

Strategies

Using soil maps, farmers can pursue two strategies to adjust field inputs:

- Predictive approach: based on analysis of static indicators (soil, resistivity, field history, etc.) during the crop cycle.
- Control approach: information from static indicators is regularly updated during the crop cycle by:

1. sampling: weighing biomass, measuring leaf chlorophyll content, weighing fruit, etc.
2. remote sensing: measuring parameters like temperature (air/soil), humidity (air/soil/leaf), wind or stem diameter is possible thanks to Wireless Sensor Networks
3. proxy-detection: in-vehicle sensors measure leaf status; this requires the farmer to drive around the entire field.
4. aerial or satellite remote sensing: multispectral imagery is acquired and processed to derive maps of crop biophysical parameters, including indicators of disease. Airborne instruments are able to measure the amount of plant cover and to distinguish between crops and weeds.

Decisions may be based on decision-support models (crop simulation models and recommendation models), but in the final analysis it is up to the farmer to decide in terms of business value and impacts on the environment.

It is important to realize why PA technology is or is not adopted, "for PA technology adoption to occur the farmer has to perceive the technology as useful and easy to use. It might be insufficient to have positive outside data on the economic benefits of PA technology as perceptions of farmers have to reflect these economic considerations."

IMPORTANCE OF PRECISION AGRICULTURE

Precision agriculture is one of many modern farming practices that make production more efficient. With precision agriculture, farmers and soils work better, not harder.

A better name for precision ag might be "site-specific ag". Growers are able to take large fields and manage them as though they are a group of small fields. This reduces the misapplication of products and increases crop and farm efficiency.

It has been said farmers were the first land stewards. They use research about weather patterns, soil temperature and humidity, growth, and other factors. They rotate crops to improve diversity, and monitor irrigation rates so that salts do not accumulate. They also use precision agriculture practices to apply nutrients, water, seed, and other agricultural inputs to grow more crops in a wide range of soil environments. Precision ag can help farmers know how much and when to apply these inputs.

There is a lot of technology used to make modern agriculture more efficient. For example, some farmers use global positioning systems (GPS) and GPS-computer guided tractors and harvestors. Other geo-referenced site-specific practices may include:

- electromagnetic soil mapping
- soil sample collection
- crop yield data collection
- aerial imagery
- crop or soil color index maps
- soil types
- soil characteristics
- drainage level
- potential yields

Each of these geo-referenced data layers helps subdivide a large field area into smaller management zones. Using small management zones reduces waste while increasing production potential.

One example of a precision agriculture practice is to evaluate the natural soil variability of a field. If the soil in one area holds water better, crops can be planted more densely and irrigation can be sparing. Or, if the plot is used for grazing, more cattle can graze than a similar area of poorer quality soil.

By studying these factors and using precision agriculture, farmers are able to produce more food at a fraction of the cost. Farmers also conserve soil for sustainable food production. Precision ag results in a stable food supply, which results in a strong community.

Implementing practices

New information and communication technologies (NICT) make field-level crop management more operational and easier to achieve for farmers. Application of crop management decisions calls for agricultural equipment that supports variable-rate technology (VRT), for example varying seed density along with variable-rate application (VRA) of nitrogen and phytosanitary products.

Precision agriculture uses technology on agricultural equipment (e.g. tractors, sprayers, harvesters, etc.):

- positioning system (e.g. GPS receivers that use satellite signals to precisely determine a position on the globe);
- geographic information systems (GIS), i.e., software that makes sense of all the available data;
- variable-rate farming equipment (seeder, spreader).

Usage Around the World

The concept of precision agriculture first emerged in the United States in the early 1980s. In 1985, researchers at the University of Minnesota varied lime inputs in crop fields. It was also at this time that the practice of grid sampling appeared (applying a fixed grid of one sample per hectare). Towards the end of the 1980s, this technique was used to derive the first input recommendation maps for fertilizers and pH corrections. The use of yield sensors developed from new technologies, combined with the advent of GPS receivers, has been gaining ground ever since. Today, such systems cover several million hectares.

In the American Midwest (US), it is associated not with sustainable agriculture but with mainstream farmers who are trying to maximize profits by spending money only in areas that require fertilizer. This practice allows the farmer to vary the rate of fertilizer across the field according to the need identified by GPS guided Grid or Zone Sampling. Fertilizer that would have been spread in areas that don't need it can be placed in areas that do, thereby optimizing its use.

Around the world, precision agriculture developed at a varying pace. Precursor nations were the United States, Canada and Australia. In Europe, the United Kingdom was the first to go down this path, followed closely by France, where it first appeared in 1997-1998. In Latin America the leading country is Argentina, where it was introduced in the middle 1990s with the support of the National Agricultural Technology Institute. Brazil established a state-owned enterprise, Embrapa, to research and develop sustainable agriculture. The development of GPS and variable-rate spreading techniques helped to anchor precision farming management practices. Today, less than 10% of France's farmers are equipped with variable-rate systems. Uptake of GPS is more widespread, but this hasn't stopped them using precision agriculture services, which supplies field-level recommendation maps.

One third of the global population still relies on agriculture for a living. Although more advanced precision farming technologies require large upfront investments, farmers in developing countries are benefitting from mobile technology. This service assists farmers with mobile payments and receipts to improve efficiencies. For example, 30,000 farmers in Tanzania use mobile phones for contracts, payments, loans, and business organization.

The economic and environmental benefits of precision agriculture have also been confirmed in China, but China is lagging behind countries such as Europe and the United States because the Chinese agricultural system is characterized by small-scale family-run farms, which makes the adoption rate of precision agriculture lower than other countries. Therefore, China is trying to better introduce precision agriculture technology into its own country and reduce some risks, paving the way for China's technology to develop precision agriculture in the future.

ECONOMIC AND ENVIRONMENTAL IMPACTS

Precision agriculture, as the name implies, means application of precise and correct amount of inputs like water, fertilizer, pesticides etc. at the correct time to the crop for increasing its productivity and maximizing its yields. Precision agriculture management practices can significantly reduce the amount of nutrient and other crop inputs used while boosting yields. Farmers thus obtain a return on their investment by saving on water, pesticide, and fertilizer costs.

The second, larger-scale benefit of targeting inputs concerns environmental impacts. Applying the right amount of chemicals in the right place and at the right time benefits crops, soils and groundwater, and thus the entire crop cycle. Consequently, precision agriculture has become a cornerstone of sustainable agriculture, since it respects crops, soils and farmers. Sustainable agriculture seeks to assure a continued supply of food within the ecological, economic and social limits required to sustain production in the long term.

A 2013 article tried to show that precision agriculture can help farmers in developing countries like India.

Precision agriculture reduces the pressure on agriculture for the environment by increasing the efficiency of machinery and putting it into use. For example, the use of remote management devices such as GPS reduces fuel consumption for agriculture, while variable rate application of nutrients or pesticides can potentially reduce the use of these inputs, thereby saving costs and reducing harmful runoff into the waterways.

WHY DID THE PRECISION AGRICULTURE IMPORTANT

A better name for precision ag might be "site-specific ag". Growers are able to take large fields and manage them as though they are a group of small fields. This reduces the misapplication of products and increases crop and farm efficiency.

It has been said farmers were the first land stewards. They use research about weather patterns, soil temperature and humidity, growth, and other factors. They rotate crops to improve diversity, and monitor irrigation rates so that salts do not accumulate. They also use precision agriculture practices to apply nutrients, water, seed, and other agricultural inputs to grow more crops in a wide range of soil environments. Precision ag can help farmers know how much and when to apply these inputs.

There is a lot of technology used to make modern agriculture more efficient. For example, some farmers use global positioning systems (GPS) and GPS-computer guided tractors and harvesters. Other geo-referenced site-specific practices may include:

- electromagnetic soil mapping
- soil sample collection
- crop yield data collection
- aerial imagery
- crop or soil color index maps
- soil types
- soil characteristics
- drainage level
- potential yields

Each of these geo-referenced data layers helps subdivide a large field area into smaller management zones. Using small management zones reduces waste while increasing production potential.

One example of a precision agriculture practice is to evaluate the natural soil variability of a field. If the soil in one area holds water better, crops can be planted more densely and irrigation can be sparing. Or, if the plot is used for grazing, more cattle can graze than a similar area of poorer quality soil.

By studying these factors and using precision agriculture, farmers are able to produce more food at a fraction of the cost. Farmers also conserve soil for sustainable food production. Precision ag results in a stable food supply, which results in a strong community.

Precision agriculture relies upon specialized equipment, software and IT services. The approach includes accessing real-time data about the conditions of the crops, soil and ambient air, along with other relevant information such as hyper-local weather predictions, labor costs and equipment availability. Predictive analyticssoftware uses the data to provide farmers with guidance about crop rotation, optimal planting times, harvesting times and soil management.

Sensors in fields measure the moisture content and temperature of the soil and surrounding air. Satellites and robotic drones provide farmers with real-time images of individual plants. Information from those images can be processed and integrated with sensor and other data to yield guidance for immediate and future decisions, such as precisely what fields to water and when or where to plant a particular crop.

Agricultural control centers integrate sensor data and imaging input with other data, providing farmers with the ability to identify fields that require treatment and determine the optimum amount of water, fertilizers and pesticides to apply. This helps the farmer avoid wasting resources and prevent run-off, ensuring that the soil has just the right amount of additives for optimum health, while also reducing costs and controlling the farm's environmental impact.

Emerging Technologies

Precision agriculture is an application of breakthrough digital farming technologies. Over $4.6 billion has been invested in agriculture tech companies—sometimes called agtech.

Robots

Self-steering tractors have existed for some time now, as John Deere equipment works like a plane on autopilot. The tractor does most of the work, with the farmer stepping in for emergencies. Technology is advancing towards driverless machinery programmed by GPS to spread fertilizer or plow land. Other innovations include a solar powered machine that identifies weeds and precisely kills them with a dose of herbicide or lasers. Agricultural robots, also known as AgBots, already exist, but advanced harvesting robots are being developed to identify ripe fruits, adjust to their shape and size, and carefully pluck them from branches.

DRONES AND SATELLITE IMAGERY

Advances in drone and satellite technology benefits precision farming because drones take high quality images, while satellites capture the bigger picture. Light aircraft pilots can combine aerial photography with data from satellite records to predict future yields based on the current level of field biomass. Aggregated images can create contour maps to track where water flows, determine variable-rate seeding, and create yield maps of areas that were more or less productive.

The Internet of Things

The Internet of things is the network of physical objects outfitted with electronics that enable data collection and aggregation. IoT comes into play with the development of sensors and farm-management software. For example, farmers can spectroscopically measure nitrogen, phosphorus, and potassium in liquid manure, which is notoriously inconsistent.They can then scan the ground to see where cows have already urinated and apply fertilizer to only the spots that need it. This cuts fertilizer use by up to 30%. Moisture sensors in the soil determine the best times to remotely water plants. The irrigation systems can be programmed to switch which side of tree trunk they water based on the plant's need and rainfall.

Innovations are not just limited to plants—they can be used for the welfare of animals. Cattle can be outfitted with internal sensors to keep track of stomach acidity and digestive problems. External sensors track movement patterns to determine the cow's health and fitness, sense physical injuries, and identify the optimal times for breeding. All this data from sensors can be aggregated and analyzed to detect trends and patterns.

As another example, monitoring technology can be used to make beekeeping more efficient. Honeybees are of significant economic value and provide a vital service to agriculture by pollinating a variety of crops. Monitoring of a honeybee colony's health via wireless temperature, humidity and CO2 sensors helps to improve the productivity of bees, and to read early warnings in the data that might threaten the very survival of an entire hive.

Smartphone Applications

Smartphone and tablet applications are becoming increasingly popular in precision agriculture. Smartphones come with many useful applications already installed, including the camera, microphone, GPS, and accelerometer. There are also applications made dedicated to various agriculture applications such as field mapping, tracking animals, obtaining weather and crop information, and more. They are easily portable, affordable, and have a high computing power.

Machine Learning

Machine learning is commonly used in conjunction with drones, robots, and internet of things devices. It allows for the input of data from each of these sources. The computer then processes this information and sends the appropriate actions back to these devices. This allows for robots to deliver the perfect amount of fertilizer or for IoT devices to provide the perfect quantity of water directly to the soil. The future of agriculture moves more toward a machine learning architecture every year. It has allowed for more efficient and precise farming with less human manpower.

In most developing countries, farming is the best safety net against hunger and keeping production costs under control without sacrificing crop yield is essential for survival. The Handbook of Precision Agriculture presents up-to-date research, field studies, and practical applications from around the world to help agricultural scientists and farmers work together to design a farming methodology that improves productivity, profitability, and sustainability. The book offers a comprehensive guide to basic principles and technologies, crop-specific applications, integrative strategies, economic and environmental concerns, and future trends in precision agriculture in different regions of the world.

Precision agriculture technology, when properly integrated into the crop production process, can greatly improve overall production and sustainability. Precision Agriculture Technology for Crop Farming focuses on the technology of site-specific crop management; detailing applications for sensing, data handling, modeling, and control. Written by experts who have contributed significantly to the development of precision agriculture technology, this book reviews its history-establishing background on the various processes and applications-describes the current status, and offers insight into the future technology of precision agriculture.

THE BASICS OF PRECISION FARMING

Precision farming enables the farmers to practice agriculture in portioned areas that make weeding and management of the crops easier. Through precision farming, the farmer will have increased harvest with reduced expenditure in managing the plants. Precision farming enables the farmers to improve their farming skills and perfect on them. With the basics of precision farming, it ensures less input with increased profit. However, in precision farming, some fundamentals must be put into consideration that will enable one to eliminate extra expenses. Below are the essential precision farming basics:

Significance of measuring

During traditional farming, farmers used to measure portions of their farms before planting for accuracy reasons. Ignoring measuring the farm will cause challenges to the farmers in managing the farm. Therefore, farmers should consider measuring as a crucial issue for the better outcome in the end. In precision farming taking measure is the basis of success towards great harvest later and easier management of the field. Measurement is the basis towards accurate farming and for better results; Precision farming in portions Precision farming can be dealt with in three parts for more natural management of the field. The three-section include; tractor and machine control, aiming agronomy and information management. This will enable the farmers to operate their precision farming more efficiently and conveniently. Successful precision farming must be based on the above three section for the better outcome. Getting in touch with machine suppliers, agronomy organization, and data management organization will enable the farmer to integrate both parts efficiently for proper precision farming.

Digital or physical farm

A connection between digital and physical farm has always been a threat towards precision farming among the farmers. This threat arises when the farm management software differs in the name of the physical farm and the digital farm. The fields names created on the management system at the different period for the reasons of taking data for the yield that differ either in the naming system or misspell of the filed name making a variation between digital and the physical farm. When the owner of the farm uploads data into the management, that is, software that figures and the names of the fields differ making management of precision farming a challenge.

As a result, if the digital farm naming differs from the physical farm changes must be made to ensure that both information is rhyming for proper management of precision farming. The farm management systems are grounded on a standard order of farmers, farms, and fields and are all the same regardless of any system available.

Poor linkage of the digital and physical naming of the fields is a challenge to precision farming which can finally hinder the progress of precision farming. Farmers should ensure that entry of digital fields and physical fields must have the same name for successful precision farming.

Guidance in line management

Proper line management is the primary factor towards successful precision farming among the farmers doing precision farming. Farmers must have a reference line A-B in their farms for operation. Having a consistent A-B management line will help in operating farm vehicles and other machines in advanced farm management techniques such as controlled traffic farming. Controlled traffic farming ensures that the farm vehicles can work in a given area at a time making precision farming effective and more manageable.

Precision Agronomy

Precision agronomy has been made possible through the use of immeasurable advanced devices, sensors, and tools available. However, another precision farming process has remained uncomplicated and made simple. The primary factor towards precision agronomy is the implementation of devices and tools that make use of agronomy knowledge in precision farming.

When the farming filed is growing large the farmer should be thinking of creating new zones or making the division within the field. The division can be done depending on the soil texture of different sectors within the filed. This is to ensure that precision farming has become more active.

The basic steps in precision farming are,

(i) Assessing variation

(ii) Managing variation and

(iii) Evaluation

The available technologies enable us in understanding the variability and by giving site specific agronomic recommendations we can manage the variability that make precision agriculture viable. And finally evaluation must be an integral part of any precision farming system. The detailed steps involved in each process are clearly depicted in a diagramme.

(i) Assessing variability

Assessing variability is the critical first step in precision farming. Since it is clear that one cannot manage what one does not know. Factors and the processes that regulate or control the crop performance in terms of yield vary in space and time. Quantifying the variability of these factors and processes and determining when and where different combinations are responsible for the spatial and temporal variation in crop yield is the challenge for precision agriculture.

Techniques for assessing spatial variability are readily available and have been applied extensively in precision agriculture. The major part of precision agriculture lies in assessing to spatial variability. Techniques for assessing temporal variability also exist but the simultaneous reporting a spatial and temporal variation is rare. We need both the spatial and temporal statistics. We can observe the variability in yield of a crop in space but we cannot predict the reasons for the variability. It needs the observations at crop growth and development over the growing season, which is nothing but the temporal variation. Hence, we need both the space and time statistics to apply the precision farming techniques. But this is not common to all the variability/factor that dictate crop yield. Some variables are more produced in space rather with time, making them more conducive to current forms of precision management.

Steps in accessing variability:

1. Surveys
2. Interpolation of Point Samples
3. High-Resolution Sensing
4. Modeling

(ii) Managing variability

Once variation is adequately assessed, farmers must match agronomic inputs to known conditions employing management recommendations. Those are site specific and use accurate applications control equipment. We can use the technology most effectively. In site-specific variability management we can use GPS instrument, so that the site specificity is pronounced and management will be easy and economical. While taking the soil/plant samples, we have to note the sample site coordinates and further we can use the same for management. This results in effective use of inputs and avoids any wastage and this is what we are looking for.

The potential for improved precision in soil fertility management combined with increased precision in application control make precise soil fertility management as attractive, but largely unproven alternative to uniform field management. For successful implementation, the concept of precision soil fertility management requires that within-field variability exists and is accurately identified and reliably interpreted, that variability influences crop yield, crop quality and for the environment. There fore inputs can be applied accurately.

The higher the spatial dependence of a manageable soil property, the higher the potential for precision management and the greater its potential value. The degree of difficulty, however, increases as the temporal component of spatial variability increases. Applying this hypothesis to soil fertility would support that Phosphorus and Potassium fertility are very conducive to precision management because temporal variability is low. For N, the temporal component of variability can be larger than its spatial component, making precision N management much more difficult in some cases.

Steps in Managing the variability:

1. Precision Soil Fertility Management
2. Precision Pest Management
3. Crop Management
4. Water Management
5. Soil Management

(iii) Evaluation

There are three important issues regarding precision agriculture evaluation.

1. Economics
2. Environment and
3. Technology transfer

The most important fact regarding the analysis of profitability of precision agriculture is that the value comes from the application of the data and not from the use of the technology. Potential improvements in environmental quality are often cited as a reason for using precision agriculture. Reduced agrochemical use, higher nutrient use efficiencies, increased efficiency of managed inputs and increased production of soils from degradation are frequently cited as potential benefits to the environment. Enabling technologies can make precision agriculture feasible, agronomic principles and decision rules can make it applicable and enhanced production efficiency or other forms of value can make it profitable.

The term technology transfer could imply that precision agriculture occurs when individuals or firms simply acquire and use the enabling technologies. While precision agriculture does involve the application of enabling technologies and agronomic principles to manage spatial and temporal variability, the key term is manage. Much of the attention in what is called technology transfer has focused on how to communicate with the farmer. These issues associated with the managerial capability of the operator, the spatial distribution of infrastructure and the compatibility of technology to individual farms will change radically as precision agriculture continues to develop (Pierce and Peter, 1999).

The First Wave of Precision Agriculture

Precision agriculture was born with the introduction of GPS guidance for tractors in the early 1990s, and the adoption of this technology is now so widespread globally that it's probably the most commonly-used example of precision ag today. John Deere was the first to introduce this technology using GPS location data from satellites. A GPS-connected controller in a farmer's tractor automatically steers the equipment based on the coordinates of a field. This reduces steering errors by drivers and therefore any overlap passes on the field. In turn, this results in less wasted seed, fertilizer, fuel, and time.

Precision agronomics is another important term related to the combining of methodology with technology. At its core, it's about providing more accurate farming techniques for planting and growing crops. Precision agronomics can involve any of the following elements:

Variable rate technology (VRT) – VRT refers to any technology that enables the variable application of inputs and allows farmers to control the amount of inputs they apply in a specific location. The basic components of this technology include a computer, software, a controller and a differential global positioning system (DGPS). There are three basic approaches to using VRT – map-based, sensor-based and manual. The adoption of variable rate technology is currently estimated at 15% in North America and is expected to continue to grow rapidly over the next five years.

GPS soil sampling – Testing a field's soil reveals available nutrients, pH level, and a range of other data that is important for making informed and profitable decisions. In essence, soil sampling allows growers to consider productivity differences within a field and formulate a plan that takes these differences into account. Collection and sampling services that are worth the effort will allow the data to be used for input for variable rate applications for optimizing seeding and fertilizer.

Computer-based applications – Computer applications can be used to create precise farm plans, field maps, crop scouting and yield maps. This, in turn, allows for the more precise application of inputs such as pesticides, herbicides, and fertilizers, thus helping to reduce expenses, produce higher yields and create a more environmentally-friendly operation. The challenge with these software systems is they sometimes deliver a narrow value that doesn't allow data to be used for making bigger farm decisions, especially with the support of an expert. Another concern with many software applications is poor user interfaces, and the inability to integrate the information they provide with other data sources to enrich and show significant value to farmers.

Remote sensing technology – Remote sensing technology has been in use in agriculture since the late 1960s. It can be an invaluable tool when it comes to monitoring and managing land, water, and other resources. It can help determine everything from what factors may be stressing a crop at a specific point in time to estimating the amount of moisture in the soil. This data enriches decision-making on the farm and can come from several sources including drones and satellites.

At its most basic level, precision agronomics takes the role of an agronomist and helps make the methods they use more accurate and scalable.

The primary aim of precision agriculture and precision agronomics is to ensure profitability, efficiency, and sustainability while protecting the environment. This is achieved by using the big data gathered by this technology to guide both immediate and future decisions on everything from where in the field to apply a particular rate, to when it's best to apply chemical, fertilizer or seed.

While precision agriculture principles have been around for more than 25 years, it's only been over the past decade that they have become mainstream due to technological advancements and the adoption of other, broader technologies. The adoption of mobile devices, access to high-speed internet, low cost and reliable satellites – for positioning and imagery — and farm equipment that's optimized for precision agriculture by the manufacturer, are some of the key technologies characterizing the trend for precision agriculture. Some experts have suggested that more than 50% of today's farmers use at least one precision farming practice.

Advocating for Excellence

Precision agriculture innovation continues, and more and more farms are adopting available technology and practices. Like any other industry, we need more advocates to drive greater adoption and hence greater efficiency. Growers need support to successfully implement new technologies to ensure success. At Decisive Farming, we support our growers with training and expertise.

TOOLS FOR IMPLEMENTATION OF PRECISION AGRICULTURE

Production issues can easily be addressed through precision agriculture if it is taken care of properly. Farmers need to identify their specific needs beforehand so that identification of ideal tools to suit their needs will follow. These tools are essential in creating efficiency, more accurate records, and ease farm management. Therefore determining your needs before spending any coin on software and gadgets is important.

After determining the needs, you can now choose the tools that you want to incorporate to solve your needs. Precision agriculture tools, software packages and guidance systems vary in prices and quality. Getting expert advice is thus very important for any guidance. Moreover, farmers should make sure that they understand what it will take to put new precision tools to work and what is expected of them when using them so as to achieve greater success in farming. Farmers should be aware if their particular tools can be used for different types of crops; or if it will be used only for a section of the farm or will it be the entire piece of land; and which of farm's staff will be handling and using the tools or if there is need for any specialized training or support will be necessary. This kind of implementation plans has significant impact on the tools you choose to meet specific needs. Having a plan for implementation on time will help farmers to enjoy maximum success with the precision agriculture tools.

Finally, it is good to be patient. Using innovative agricultural tools and be comfortable with them might take time. Do not be frustrated. It may take more than one season to fully implement your precision tools. This is because you might be trying to work out the bugs while trying to completely integrate and implement the system o that you may get the desired achievements you intended to have.

Precision Tools

Auto-guidance equipment

This type of technology allows farmers to cover a vast field accurately and faster. It essentially eliminates overlaps, skips and gaps from any product that farmers apply. As a result, fertilizer, insecticides, pesticides and other crop protection products are only applied at the right amount and places that are in need. Time is money and thus auto-guidance tools saves on both. Auto-guidance tools also minimises accidents that are usually caused by these machines.

Variable-Rate Technology

Variable rate technology (VRT) fertilizer, insecticide, water and pesticides applications can be beneficial in fighting pests, diseases and balancing soil pH levels that might be either too high or too low in different regions of the same field. VRT applications of potassium (K) and phosphorus (P) can be beneficial, depending on the fertility of the field. Management zones should be developed rather than relying on grid soil samples alone.

Proximate sensors technology

There are different types of sensors. Moisture and temperature sensors are some of them. These sensors are able to help the farmer to detect any problem experienced by the crops, or stress that livestock might be undergoing. It then helps the farmer fix the problems easily with fewer resources.

Global positioning system and Geographical information system

GPS is a type of satellite that identifies the location of field equipment which is within a meter of the actual site in the farm. GIS consists of software that input, retrieves, stores, displays and analyses referenced geographical information in map like form. The importance of knowing a precision location within few inches is that:

- The laboratory results and location of soil samples can easily be compared to a soil map.
- Pesticides and fertilizer can be prescribed to fit the specific soil properties
- Tillage can be adjusted because one condition varies across the field.
- From monitoring farmers can record yield data while on the move across the field.

Top 10 technologies

1. GPS/GNSS

It's hard to tell exactly where the state of precision agriculture today would be without GPS — literally. From virtually the moment agriculture gained access to position locating satellites in the 1990s, operators and manufacturers have found various ways to tie into these tools to make managing field work much easier and accurate. "In North America and Europe, growers can turn on the tractor and get to work almost immediately," says T.J. Schulte, Marketing Manager for Trimble Agriculture Division.

Looking beyond these capabilities, experts say that satellite technology is truly deserving of its "global" moniker. "No longer can we refer to all these systems as GPS — that's not an accurate description when referring to new Global Navigation Satellite Systems (GNSS) receiver technology today," says Greg Guyette, President of Insero. Instead, he adds, GNSS covers all countries' satellite constellations including GPS, GLONASS, and Galileo.

2. Mobile Devices

After figuring out where precision agriculture stands on the planet, the next most important innovation these past 20 years would have to be the development of mobile devices. The world today would be an entirely different place without them, according to Illinois Grower John Reifsteck. "Without the cell phone, we probably would still be sitting in the barn a lot, waiting for someone to come to the barn and make things work," says Reifsteck.

Today, cell phones have morphed into a whole host of useful mobile devices including smartphones and tablets. So ingrained has this technology become that experts estimate that there are more mobile devices on today — 7.25 billion — than people (around 7.2 billion).

3. Robotics

Robots are taking on many tasks in agriculture these days (with varying levels of success), including planting greenhouse crops and pruning vineyards. And there's been no shortage of activity in agronomic crops. The biggest push has been for autonomous machines that are remotely controlled using telematics. Kinze engineers have created an autonomous grain cart system (designed to plug into any tractor) in which the cart follows a combine through the field at safe distance.

Launched in 2011, AGCO's Fendt Guide Connect leader-follower technology also connects two machines by means of GNSS signal and radio, so that both can be controlled by just one driver. AGCO is continuing to develop the concept based on customers' input on their farming needs, says Sepp Nuscheler, Fendt Senior Communications Manager at AGCO.

In a different approach, the Fendt MARS (Mobile Agricultural Robot Swarms) project utilizes small corn seeding robots that are lightweight, energy-efficient, highly agile, cloud-controlled and operated from a tablet app. There's no cab but one off-field operator managing a fleet of multiple MARS units. They can work around the clock and have low maintenance needs. "Look for some exciting developments to be shared on the MARS project in Q4 of this year," says Nuscheler. "This is one direction we see the future of ag robotics heading — small but smart and many."

4. Irrigation

Innovations in precision irrigation technologies are growing even more crucial as growers face water scarcity due to drought, aquifer depletion, and water allocations. One recent advance is telemetry, says John Campbell, Manager of Technology Advancement and Adoption with Valley Irrigation. Products now allow growers to remotely monitor and control nearly every facet of their irrigation operation. The systems save water, time, fuel, and wear and tear on vehicles.

In the future, Campbell says producers will be integrating soil moisture monitoring, weather data and variable-rate irrigation (VRI) into their systems.

5. Internet of Things

One of the newest buzzwords to hit precision over the past few of years is the "Internet of Things" (IoT). Simply defined, it's the concept of connecting any device with an on/off switch to the Internet (and/or to each other). This network of connected things could also include people with wearable devices.

The idea has been demonstrated in the consumer market in the "connected home," for instance, where appliances, security systems, and the like communicate with each other and the homeowner. Craig Houin, Data Management Lead at Sunrise Cooperative, says connected components in agriculture could include field sensors (for logging real-time weather, soil moisture, and temperature data) and aerial/satellite imagery for field monitoring. Such device communications could also be used in dispatching programs, sales interaction tools, and other business management applications.

Most recently, a number of ag start-ups and component suppliers (hardware, software, etc.) are using LPWANs (Low Power Wide Area Network) in place of or to augment cellular networks in wireless data transmission. "These networks are designed to carry small amounts of data transmitted intermittently over long ranges," says Paul Welbig, Director of Business Development at Senet Inc. Because the devices that communicate with the LPWA networks do so with very low power, their battery lives are substantially longer than the current cellular offerings. This coupled with low cost network usage provides a very compelling total cost of ownership advantage over other options.

6. Sensors

Wireless sensors have been used in precision ag and/to gather data on soil water availability, soil compaction, soil fertility, leaf temperature, leaf area index, plant water status, local climate data, insect-disease-weed infestation, and more. Perhaps the most advanced and diverse technologies to date are found in water management. Across the country, increased regulation of water use and water scarcity will continue to drive improvements in this area. In fact, BCA Ag Technologies' Ben Flansburg says soil moisture and rainfall sensors have been some big sellers. And many more producers in California are using moisture sensors to help irrigation scheduling.

On-the-go sensor information has become more valuable as well. On-board applicator options developed over the past few years include GreenSeeker (Trimble), OptRx (Ag Leader), and CropSpec (Topcon). They communicate real-time crop health conditions to help immediately tailor product applications.

Another innovation? WeedSeeker, Trimble's weed detection sensor made for precise site-specific application of herbicides. "Its use is growing in geographic regions where weeds have developed a tolerance to standard glyphosate-based broad-spectrum herbicides," notes Mike Martinez, Marketing Director.

7. Variable Rate Seeding

Given all the newer/exciting technologies for precision agriculture on this list, it might be a surprise to see variable-rate application (VRA) seeding here. According to Sid Parks, Manager of Precision Farming for GROWMARK, this has been able to maintain its importance in part because of its nature. "It appeals to a growers' natural inclination to try to maximize a field to take advantage of all of the possibilities of increasing the yields possible by paying extra attention to the factors that impact seed growth," says Parks. "It's a little different than variable-rate fertilizer because VRA seeding relies on your ability to gather accurate data for the start of the agricultural process, the seed itself."

8. Weather Modeling

Visit most of the nation's ag retail locations and chances are some kind of weather tracker will be on display. Perhaps no other variable is as important — and completely unpredictable — as the weather.

But help is on the way. "Over the past 25 years, you've gotten a lot of important technologies developed that would be even more valuable with quality weather modeling, but we are now at a point where assuring good crop yields is key to making certain the world has food solutions that work," says Jeff Keiser, Vice President of Strategic Sales and Marketing for Iteris. "Here at Iteris, we've been in the data modeling business for more than 30 years. Our first agricultural system, ClearAg creates a platform for agriculture and expands into other modeling areas such as water use, soil properties, and crop growth."

An example of this, says Keiser, involved a potato grower in the Northern Plains that found harvesting his crop at a certain temperature was key for maintaining crop quality and integrity. In the past, this grower sent scouts out into the field to manually assess soil temperatures before sending in the harvest equipment. "But by using ClearAg, this grower was able to take all their soil readings remotely and he was able to accomplish his harvest a lot more efficiently than ever before," he says.

9. Nitrogen Modeling

Although some forms of variable-rate fertilizer have been used for decades, nitrogen modeling has become more pronounced recently. "Our clientele has been using variable-rate technologies for fertilizer applications since the mid-1990s," says Matt Waits, CEO for SST Software. "However, the complexity of the nitrogen cycle and how it is in a constant state of flux has always made managing nitrogen difficult."

10. Standardization

The call for compatibility across equipment manufacturers' components — primarily through ISOBUS standards — continues to go out. Official initial efforts to implement this began about eight years ago with the formation of the Agricultural Industry Electronics Foundation. The group now includes more than 170 companies, associations, and organizations that are actively collaborating to make the standards work. Industry participants that have to deal with equipment compatibility "on the ground" continue to be frustrated, however. Third-party tech experts often struggle to manage competing suppliers' products. Says Tim Norris, CEO of Ag Info Tech, LLC, Mount Vernon, OH: "There will be a point hopefully where components get to be pretty much plug-and-play — and it's a lot better than it was — but it's still a real issue."

Once ranking based on variability has been done, look for fields that would require fewer inputs with the intention of saving money. These areas are the less productive zones. What follows then is addressing the underperforming areas. Minimising inputs in low productivity areas will result to reduced cost of farming and in the long run increased revenues during harvesting season. Get the correct limiting factors of yield in the zones with higher productivity. Such attributes are but not limited to; soil texture, pH value, water table, nutrient imbalance, weed patches, soil profile depth, and soil compaction among many other factors. This will help you adjust VR rate and your yield goals. Further, you may set your target goals in the high productivity zones in an aggressive manner, more than you set in other regions. You can then adjust your fertilizer application rate to meet the nutritional value of your farm. Finally you can measure the returns and the results. These measurements are good in determining farm best practises. At a glance you will be able to see whether your farm is paying off and areas where amendments and cost of inputs are not paying off. Solutions such Trimble Ag gives profit maps to farmers as well as VR against fertilizer rate.

The next step is to establish the kind of tools that are critical in meeting your above needs. Determining the right tools is the first section; this is because shopping for the right tools, models and methods can be challenging. Precision tools that seem to be functionally similar may vary in quality and prices. You may consult in order to get guidance regarding software systems and machinery capabilities and determine which tool I right for you.

Next is understanding the requirements needed to implement the new precision tool that you are going to use with maximum success. Questions on whether that tool could be used for multiple crops, who will run it, the knowledge required to deploy it and its coverage per piece of a farm should be considered so as to get success in precision farming technology.

Lastly, be patient. Adding a whole new technology to farm may take a while to adapt. New farming operations may require slight adjustments. You will be require to readjust your adoption timelines from time to time to fully get the system needed to obtain the desired results.

In summary, the following are advantages to farmers:

- Overall increase in yield. Precise selection of areas in need and application of various farm inputs ensure optimal results are attained.

- Efficiency improvement. Advance technologies reduces wastage and saves on labour costs
- Reduced cost of production. Application of exact quantities and at the right time reduces wastage.
- Results into better decision making
- Reduced environmental impact.

Conclusion

Successful implementation of precision agriculture is not a quick fix thing, but rather, it is usually an evolving process that might not show immediate results. You may need multiple methodologies or tools to address some concerns. Do not hesitate to consult and explore the many existing precision agriculture opportunities and technologies.

TE EMERGENCE AND CURRENT STATUS OF PRECISION AGRICULTURE

Present Status of Precision Farming: A Review

Precision Agriculture is generally defined as information and technology based farm management system to identify, analyze and manage variability within fields for optimum profitability, sustainability and protection of the land resource. In this mode of farming, new information technologies can be used to make better decisions about many aspects of crop production. Precision agriculture highlights the increased efficiency that can be realized by understanding and dealing with the natural variability found within a field. The goal is not to obtain the same yield everywhere, but rather to manage and distribute inputs on a site specific basis to maximize long term cost/benefit. input costs and decreasing commodity prices, the farmers are looking for new ways to increase efficiency and cut costs. Precision farming technology would be a viable alternative to improve profitability and productivity. Applying the same inputs across the entire field may no longer be the best choice . Today because of increasing input costs and decreasing commodity prices, the farmers are looking for new ways to increase efficiency and cut costs. Precision farming technology would be a viable alternative to improve profitability and productivity.

Need for Precision Agriculture

The potential of precision farming for economical and environmental benefits could be visualized through reduced use of water, fertilizers, herbicides and pesticides besides the farm equipments. Instead of managing an entire field based upon some hypothetical average condition, which may not exist anywhere in the field, a precision farming approach recognizes site-specific differences within fields and adjusts management actions accordingly . Farmers usually are aware that their fields have variable yields across the landscape. These variations can be traced to management practices, soil properties and/or environmental characteristics. A farmer's mental information database about how to treat different areas in a field required years of observation and implementation through trial-and error.

Smart farming (SF), based on the incorporation of information and communication technologies into machinery, equipment, and sensors in agricultural production systems, allows a large volume of data and information to be generated with progressive insertion of automation into the process. Smart farming relies on data transmission and the concentration of data in remote storage systems to enable the combination and analysis of various farm data for decision making.

Today, that level of knowledge of field conditions is difficult to maintain because of the larger farm sizes and changes in areas farmed due to annual shifts in leasing arrangements. Precision agriculture offers the potential to automate and simplify the collection and analysis of information. It allows management decisions to be made and quickly implemented on small areas within larger fields.

Wireless Sensors in Precision Agriculture

The development of WST applications in precision agriculture makes possible to increase efficiencies, productivity and profitability while minimizing unintended impacts on wildlife and the environment, in many agricultural production systems. The real time information from the fields will provide a solid base for farmers to adjust strategies at any time.

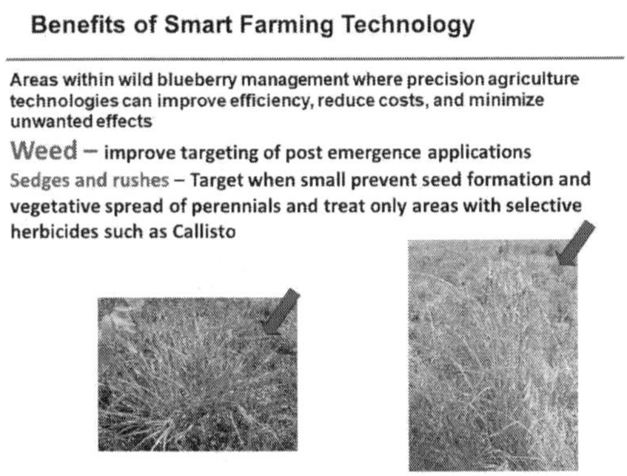

Fig 23.2

Wireless sensors have been used in precision agriculture to assist in

(1) Spatial data collection

(2) Variable-rate technology

(3) Supplying data to farmers.

Spatial data collection

This system consists of a data collection vehicle, a manager vehicle,data acquisition and control systems on farm machines. It is able to conduct local field survey and collect data of soilwater availability, soil compaction, soil fertility, biomass yield, leaf area index, leaf temperature, leaf chlorophyll content, plant water status, local climate data, insect-disease-weed infestation, grain yield, etc. The data collection vehicle retrieves data from farm machines via a WLAN, analyze, store and transmit the data to the manager vehicle wirelessly .

Variable-rate technology

Variable rate technology is being used in the development of an automated fertilizer applicator for tree crops. The system consists of an input module and real-time sensor data acquisition, a decision module for calculating the optimal quantity and spread pattern for a fertilizer, and an output module to regulate the fertilizer application rate. Data communications among the modules are established using a bluetooth network .

Supplying data to farmers

It involves a web server which provides information on pest, disease infestation and weather forecasts. Farmers can download the information directly via Internet and use them for operation scheduling .

Geographically referenced data can be displayed in the GIS, adding a visual perspective for interpretation. In addition to data storage and display, the GIS can be used to evaluate present and alternative management by combining and manipulating data layers to produce an analysis of management scenarios.

1. **Strategies**-Precision farming is still only a concept in many developing countries and strategic support from the public and private sectors is essential to promote its rapid adoption. Successful adoption, however, comprises at least three phases including exploration, analysis and execution.

2. **Outlook**-Precision agriculture can address both economic and environmental issues that surround production agriculture today. Questions remain about cost-effectiveness and the most effective ways to use the technological tools we now have, but the concept of "doing the right thing in the right place at the right time" has a strong intuitive appeal.. In the light of today's urgent need, there should be an all-out effort to use new technological inputs to make the 'Green Revolution' as an 'Evergreen Revolution'.

Ultimately, the success of precision agriculture depends largely on how well and how quickly the knowledge needed to guide the new technologies can be found.

Demographic trends, including aging populations and continued migration of people from rural to urban areas, have attracted the attention of researchers, because labor issues may become a scarcity factor in agriculture. In addition to these trends, the intensification of climate change will continue to alter growing conditions, such as the temperature, precipitation, and soil moisture, in less predictable ways . SF tools can help reduce these impacts, keep them constant or reduce production costs in agricultural activities, and they can assist in minimizing environmental constraints.

Humankind has witnessed great progress, first with the industrial revolution, healthcare, information technology, followed by biotechnology, and more recently the big data revolution. Increased life expectancy, growing population and shrinking arable land have contributed to mounting pressure on traditional agriculture (and somewhat led to dispelling the agrarian myth).

The adoption of technology and extensive potential application of information technology to agriculture has led to the evolution of agriculture into Precision Agriculture.

Precision Agriculture may be defined as a set of technologies that help farmers to adopt farm operations to manage field's variability. The goal is to optimize yield (having better nutrient content) and maximize return on investment.

Often Precision Agriculture involves satellite farming or site specific crop management based on observing, measuring, and responding to inter and intra field variability in crops.

Steps in Precision Agriculture:

The Precision Agriculture cycle can be considered to consist of the following steps or processes:

1. Gather (Data Acquisition)
2. Analyze (Processing the data and transform it into relevant analyzable information)
3. Decide (Analysis of this data for decision making, sometimes via the agronomic model of utilization)
4. Execute (Implementation/Adaptation)
5. Approaches in Precision Agriculture:
6. Broadly, Precision Agriculture can be practiced in either a Predictive Approach or a Control Approach.

The Predictive Approach is when practices are adapted prior to start of the season. The predictive approach steps can be categorized as:

7. 1. Gather—Use historical data, such as weather data, soil data, crop status during a period; as the starting inputs.
8. 2. Analyze—Characterize the agro-climatic context, define management zones.
9. 3. Decide—Adjust the crop inputs, such as seeds and fertilizer source.
10. 4. Execute —Vary the sowing density of seeds; vary the Nitrogen application within the fields.

The Control Approach is when adjustments are made during the season with an aim of optimizing the yield by reducing/optimizing the in-season requirements. The steps in control approach can be:

11. 1. Gather—Monitor crop growth conditions during the cycle.
12. 2. Analyze —Assess current intra-field variability.
13. 3. Decide—Adjust the crop inputs to crop needs.
14. 4. Execute —Vary the Nitrogen application and chemical application within the field.
15. Tools to Gather Data

TOOLS TO ANALYZE DATA

Based on the approach and the type of data gathered for the purpose of Precision Agriculture, processing of the data and transformation into relevant information can be done with the help of information technology and big data processing tools.

Tools to Decide

On the basis of the inputs from analysis step various models for decision making can be applied. For example, an averaging out approach may aim at averaging the yield throughout the field by reducing/controlling the field variability, this can be achieved by applying or pointing more resources, to regions where the growth lags behind the average field growth. The opposite approach would be to direct more resources (water, fertilizer, chemicals) to regions which already show better growth/performance compared to the field average.

Tools to Execute

Once the decision has been taken, various tools that can be adapted to meet the goals can be summarized as below. This can be done with the help of tractor combines, robotics or manual intervention. Precision Agriculture Technology is the process of utilizing technology for producing high-end equipments used on farms to enhance traditional practice of farming.

The Chief Minister of Maharashtra, Mr. Devendra Phadnavis, in the first workshop of the Centre for the Fourth Industrial Revolution India, predicted that Artificial Intelligence (AI) and the adoption of drone technology will change the face of precision agriculture in India.

Precision Agriculture is a process of farming management that leverages technology to ensure that the soil and crop receive the accurate amount of nurturing required for their productivity. Since, this farming management technique is based on exact specifications and precision of the location, it is also known as satellite-farming or site-specific crop management (SSCM).

In a nutshell, the entire process of farming is more smart, controlled and targeted, instead of being generic.

How is Disruptive Technology changing Precision Agriculture?

Artificial Intelligence, Robotics, Machine Learning and Deep Learning are changing the face of agriculture in India. In past two decades, Drones are being widely considered as a game changer in areas such as Agriculture. These aerial vehicles have the latest sensors and technology to tack and improve everything from irrigation, to the amount of fertilizers required and as well as the spacing in between crops to achieve the highest yield.

(1) Assessing health of crops: Drones are particularly useful in maintaining the health of the crops. The Multispectral imaging technology enables farmers to be take a pre-emptive and take preventive approach towards the health of their crops.

Drones can produce colour-coded images that indicates the presence and state of vegetation as well as detection of disease in plants.

(2) Speeding up the Seeding process: Technology help farmers to sow seeds to a farm in an accurate manner and in fast speed. This is particularly useful for farms that are on hill slopes and is difficult to reach. A "firing drone" uses a particular algorithm to carry out the planting strategy. The drone flies as high as six feet above the ground, shooting germinated seed pods at a high speed so that it penetrates the soil along with the necessary plant nutrients.

(3) Overall monitoring of the Farms: Now farmers can monitor their whole farm regardless of its size from a remote location. This is possible because drones are equipped with standard video and still cameras. Not just that the thermal imaging cameras on the drones indicate hot and cool areas which can be used to detect plant disease and to monitor irrigation needs. The image analysis is done through Deep Learning which is capable of advanced image recognition technology.

(4) Weed Detection through analytics: Weeds are a major destructive force for crops regardless of the farm size. Till now farmers have been manually detecting and scouting for weeds but now the work has been eased by Drones.

With the help of advanced data analytics, weed detection can be done by drones using Weed Pressure Analysis tools. The Weed Pressure algorithm analyses the HD drone data to create a Weed Pressure Analysis tool. This depicts the percentage of weeds spread across the farm.

(5) Livestock Management: Livestock farmers can use drones to monitor the area where livestock is kept. With the help of Thermal Imaging and High Definition camera, Drones can help farmers to keep an eye on their livestock from any location. The safety of the livestock, its movements and its health condition can all be accessed through the advanced imaging and analytics.

(6) Plucking fruits: The drone also helps to check the number of fruits that are suitable enough to be plucked from each tree. Every farmer know the number of fruits in each tree and the precise location where the fruits are ready for the harvest. The drone facility saves the time of farmers and improves the efficiency in harvesting.

Technology is an indispensable part of modern farming

Drones are definitely a crucial invention and a long-term solution. However, it demands close co-operation amongst governments, technology giants, and the agriculture sector. Precision Agriculture Technology should aim at optimum production, efficient management, cost-effective methods and overall sustainability. This calls for producers to adopt revolutionary strategies for increasing production in the most cost and time-efficient way.

The Time Is Right for Precision Agriculture Technologies

Precision agriculture technologies have not always been economical for small- to medium-sized farming operations.However, with precision agriculture equipment becoming less expensive, tools such as guidance systems, yield monitors and variable-rate fertilizer applicators may now be profitable for nearly all growers. The rising costs of inputs and higher commodity prices considerably increase the risk of making the wrong management decision. Thus, even small farms can profit from using technologies that improve production efficiency.

Some technologies, like real-time kinematic (RTK) auto-steering, can improve efficiency without changing management practices.Using a GPS-guided steering system can eliminate sprayer overlaps and planter skips that can result in lower profits. The magnitude of savings depends on how well the grower was doing without the guidance system. Considering an example from an Ohio State University Extension article, a grower using row or foam markers on the planter and sprayer, conservatively speaking, might be farming 102 acres in a 100-acre field. This extra area might seem insignificant, but when one considers that two acres translates into spending 2 percent more on all associated inputs such as seed, fertilizer, pesticides, fuel and labor, even small application errors can become quite costly. An RTK guidance system with 1-inch driving accuracy can eliminate this risk.

Despite the fact that yield monitors have been around over a decade, many growers still don't fully understand how to use them to improve farming efficiency. This lack of knowledge is being actively addressed in a series of extension programs and classroom courses developed at North Carolina State University. This training involves on-farm demonstrations, hands-on classroom training using "Virtual Yield Monitor" custom software, and introduction to spreadsheet-based analysis of yield monitor data, yield-limiting factors, and potential changes in management that could increase yield. Efforts like this provide growers the knowledge needed to fully utilize yield monitor technologies to better manage on-farm spatial variability.

Variable-rate fertilizer applications have been shown to improve efficiency and increase profits in many growers' fields. Several universities and USDA-ARS research units have developed strategies for using on-the-go, sensor-based applicators to improve fertilizer-use efficiency. Profits have come in the form of increased grain yields without increasing total nutrient inputs or as sustained production at lower input levels. Most of these systems consider both spatial and temporal variability, which can affect production. Current work is focused on incorporating additional layers of data such as real-time weather, soil electrical conductivity, and other spatial information into the processes that determine fertilizer application rates. Precision management pays more now than ever.Current grain and fertilizer prices greatly increase the value of precision agriculture technologies. The information generated using precision agriculture equipment and the decisions based upon it can help mitigate the growing risk of yield loss.

PRECISION AGRICULTURE: CROP CONDITION

One of the main advantages of remote-sensing is that synoptic weed data can be acquired virtually instantaneously (within the field of view of the sensor), and a weed map generated within hours of data acquisition. However, because little information is available concerning the scale at which weeds should be managed within fields, the sensing and mapping technology has tended to dictate the resolution at which weeds must be mapped. This paper summarizes the work completed to date to investigate the use of airborne remote-sensing for weed mapping in crops, and discusses application of the technology in precision weed management practices.

Scope of precision farming in India

The concept of precision farming is not new for India. Farmers try their best to do the things for getting maximum possible yield with information and technologies available to them but unless & until total information about their fields and advanced technologies are available, they cannot do precision farming in perfect sense. In India, major problem is the small field size. More than 58 percent of operational holdings in the country have size less than 1ha. Only in the states of Punjab, Rajasthan, Haryana and Gujarat more than 20 per cent of agricultural lands have operational holding size of more than four hectare. When contiguous fields with the same crop are considered, it is possible to obtain fields of over 15 ha extent in which similar crop management are followed. Such fields can be considered for the purpose of initiating the implementation of precision farming. Similar implementation can also be carried out on the state farms. There is a scope of implementing precision agriculture for crops like, rice and wheat especially in the states of Punjab and Haryana. Commercial as well as horticultural crops also show a wider scope for precision agriculture in the cooperative farms. Nearly two-third arable land in India is rain-fed. The crop yields are very low (H"1t ha) and very good potential exists for increasing productivity of rain-fed Cropping systems.

Fig 23.3

Benefits of precision farming

- The concept of "doing the right thing in the right place at the right time" has a strong intuitive appeal which gives farmers the ability to use all operations and crop inputs more effectively.
- More effective use of inputs results in greater crop yield and/or quality, without polluting the environment.
- Precision agriculture can address both economic and environmental issues that surround production agriculture today.

Drawbacks of precision farming

- High cost: It has proven difficult to determine the cost benefits of precision agriculture management. At present, many of the technologies used are in their infancy, and pricing of equipment and services is hard to pin down.
- Lack of technical expertise knowledge and technology: The success of precision agriculture depends largely on how well and how quickly the knowledge needed to guide the new technologies can be found (India spends only 0.3% of its agricultural Gross Domestic Product in Research and Development)
- Not applicable or difficult/costly for small land holdings
- Heterogeneity of cropping systems and market imperfections

The policy approach to promote precision farming at farm level:

- Identify the niche areas for the promotion of crop specific precision farming.
- Creation of multidisciplinary teams involving agricultural scientists in various fields, engineers, manufacturers and economists to study the overall scope of precision agriculture.
- Promote the progressive farmers for precision farming technology who have sufficient risk bearing capacity.

- Encourage the farmers to study of spatial and temporal variability of the input parameters using primary data at field level.
- Provide complete technical backup support to the farmers to develop pilots or models, which can be replicated on a large scale.
- Pilot study should be conducted on farmer's field to show the results of precision agriculture implementation.
- Encourage the farmers to adopt water accounting protocols at farm level and to use of micro level irrigation systems and water saving techniques.
- Government legislation restraining farmers using indiscriminate farm inputs and thereby causing ecological/environmental imbalance would induce the farmer to go for alternative approach.
- Creating awareness amongst farmers about consequences of applying imbalanced doses of farm inputs like irrigation, fertilizers, insecticides and pesticides.
- Policy support on procurement prices, efficient transfer of technology to the farmers, formulation of cooperative groups or self help groups since many of the precision agriculture tools are costly (GIS, GPS, RS, etc.).

GRAIN QUALITY

Precision Agriculture (PA) is an approach to managing the variability in production agriculture in a more economic and environmentally efficient manner. It has been pioneered as a management tool in the grains industry, and while its development and uptake continues to grow amongst grain farmers worldwide, a broad range of other cropping industries have embraced the concept This book explains general PA theory, identifies and describes essential tools and techniques, and includes practical examples from the grains industry. Readers will gain an understanding of the magnitude, spatial scale and seasonality of measurable variability in soil attributes, plant growth and environmental conditions. They will be introduced to the role of sensing systems in measuring crop, soil and environment variability, and discover how this variability may have a significant impact on crop production systems.

PRECISION AGRICULTURE FOR GRAIN PRODUCTION

Systems will empower crop and soil science students, agronomy and agricultural engineering students, as well as agronomic advisors and farmers to critically analyse the impact of observed variation in resources on crop production and management decisions.

Grain storage: maintaining grain quality

Grain stored on farms for seed or livestock feed is an important investment. Protecting the grain from insect attack can be regarded as an exercise in income protection.

Even light infestations of weevils can drastically reduce the germination of grain. Weevils contaminate and eat food intended for your livestock. Grain can lose a lot of weight through a heavy infestation. Weevils which survive over-winter in stored grain are ready to re-infest your harvesting machinery and next year's crop, leading to the risk of your new season's grain being rejected by the buyer.

Protecting the grain from moisture preserves the value of the grain. Cereals in store are a community of living organisms made up of the grain itself, insects, mites, bacteria and fungi. While the grain stays dry and cool, most of these living organisms remain in a suspended state or at a low level of activity. If the grain reaches moisture levels higher than the recommended 12% these organisms become active. Bacteria invade the grain, fungal growth proliferates, insects and mites breed more rapidly and produce even more water, all of which ruin your investment.

Moisture levels can be controlled by ensuring:

- the grain is loaded at below 12% moisture content
- insects are controlled at loading to prevent them creating moisture
- an aeration system is fitted to manipulate the internal temperature.

How to protect farm stored grain

The first essential is to practice good farm hygiene to reduce the overall numbers of insects on your property. This lowers the risk of infested grain being delivered for export. It also gives any applied treatments a better chance to work.

Before harvest - clean up

Grain handling equipment such as trucks, harvesters, augers and seeders must be cleaned after use. Remove all residues and then apply an insecticide to the clean surfaces as insects can survive even if insecticides are applied on top of the residue they are living in. Treat the cleaned surfaces with insecticide such as Dryacide®. This product will give at least 12 months insect control provided it is applied properly.

To protect your header measure out 2-2.5kg of Dryacide® and apply half to all the openings and inspection holes on the header. Run the machine to disperse this material and then apply the rest of the dust to the same apertures.

SILO STORAGE

Sealed silos

When buying a new grain silo, a factory-sealed model is a good investment in grain protection. Fumigation with phosphine generating tablets in sealed silos is one of the best ways to control weevils in farm stored grain. At least one sealed silo on each farm is essential as it provides a storage within which weevil infested grain can be fumigated before transporting off the property.

Sealed silos not only keep the gas in; they also keep the insects out, however;

- sealed silos do not control insects without the help of a fumigant
- it is essential to fumigate the silo every time grain is added.

Once the silo has been purchased, insect control is simple provided the silo is working correctly. The hatches must be sealed properly and the pressure relief valve filled with oil to the correct level. An annual check is essential.

Fumigation of a poorly maintained silo does not control insects effectively as gas will be lost, insects will survive and they may produce future populations with a degree of phosphine resistance.

Unsealed silos

Proper fumigation is impossible in unsealed silos. There may be an illusion of success when the adult insects are killed but once the gas has dissipated the next generation of insects will emerge from eggs or pupae. Existing unsealed silos may be modified and sealed but the cost varies depending on the design.

Emergency fumigation

Fumigation is most convenient and effective in a solid sealed structure. However, in the absence of such a grain store, a good fumigation can be carried out under a gasproof tarpaulin. The aim of the procedure is to create a leak-proof envelope of plastic sheeting. Standard black 'builders' plastic sheeting is not gasproof, is difficult to make leak-proof and is easily damaged. Heavy duty gasproof tarpaulins are recommended for the task.

Lay a tarpaulin on the ground and heap the grain on top, cover with a gas-proof sheet and roll the edges together. Lock them together with soil or weights to prevent gas loss. Calculate the size of the heap and apply five grams of phosphine per cubic metre.

Unroll a small section to allow the introduction of phosphine in a bag chain or sachet formulation. If tablets are used on a tray, you must create 30cm of airspace above the tray of tablets to avoid the development of explosive concentrations of gas. Allow 10 days for the gas to release and full circulation to occur. Open the rolled edges and let the remaining gas escape. Remove the tarpaulin and allow three days of ventilation before delivery.

If the tarpaulin is large enough it may be possible to dump the grain in the middle and roll the edges together on top of the pile.

The best location to evaluate productivity levels, by measuring yield and quality of grain and straw, is the combine harvester. Moreover, other grain quality characteristics like density or test weight can be determined for use as an evaluation tool. In this paper, an overview is given of the past and current research toward the evaluation of currently available commercial sensors (e.g., for measuring grain yield and grain moisture content) as well as toward the development of new sensors (e.g., grain protein content and straw yield).

White silos

Painting the silo white after installation on the pad is a useful management tool. A white painted silo can be 4°C cooler than a weathered galvanised unit.

Cooler grain temperatures have several advantages:

- seed germination percentage is maintained longer
- moisture migration is reduced
- insect breeding is reduced
- less gas is lost through the pressure relief valve of sealed silos.

Aeration

For longer term storage, aeration fans are run for a few hours when needed. The amount of cooling achieved during storage depends on the moisture content of the grain and the humidity and temperature of the incoming air. Seek advice on fan operation times from the equipment manufacturer.

Aerated silos are fitted with fans that pass controlled amounts of air through the grain. When aerated correctly, the grain is cooled and the grain temperature and moisture are equalised throughout the silo.

Fans are run for extended periods during the first few days of storage. An average of 12 hours using the coolest and driest air of the day is common. This removes harvest heat and evens out temperature and moisture, avoiding hot spots and moisture damage.

Underground storage

Low cost but makeshift underground storage has a place in some farming systems but a good natural site is essential and recovery may be difficult. This system has a place mainly for long term storage as a drought reserve. Grain left in properly designed underground stores will last for several years without any deterioration. It has even been delivered to bulk store after four years underground.

Aeration advantages

- Lower temperatures allow moist grain to be stored safely for longer periods.
- Even temperature of the grain mass in the silo prevents moisture migration.
- Hot spots are prevented from developing.
- Mould growth is slowed and insect development is reduced dramatically.
- Cool stored seed retains its viability and vigour for a longer period.

Western Australian trials

WA trials have shown that correctly controlled aeration can reduce grain temperature to 20°C or lower through summer. Use dry, cold air so that moisture levels are not increased and fit a controller that responds to temperature changes. A time proportioning controller automatically resets its thermostat to draw in only the coldest air.

Should you use aeration?

Before deciding to fit an aeration unit, it is important to consider the final destination of the grain. If the grain is to be fed out or used for seed on the property, aeration will keep insect damage at a low level by slowing reproduction. If the product is to be transported off the property for sale, it must not contain any live grain insects. Aeration alone does not guarantee insect free grain, so fumigation may also have to be carried out.

Aeratable sealable silos

A silo can be constructed as a sealed unit with an aeration system fitted to its base. A vent is installed on the top of the silo to allow the air that is forced in at the base to escape. This can be a waterproof perforated lid in place of the sealed lid or a venting tube that can be capped for fumigation.

Your silo or aeration equipment supplier should provide detailed operating instructions to help you store the grain safely in your silo.

INSECT CONTROL

Fumigation

Grain in a well maintained sealed silo may be protected indefinitely by adding two phosphine generating tablets per tonne of silo capacity. Place them on the surface of the grain in trays and observe the correct fumigation period of at least seven days.

Although a well-sealed silo will not allow insects to reenter, it is still necessary to check the grain at regular intervals for insect infestation after fumigation.

Phosphine leaves minimal residue when applied correctly and treated grain may be delivered at the bin with no penalty but a declaration must be made. It is illegal to transport grain under fumigation or deliver grain with partly decomposed phosphine tablets. Phosphine will not damage germination at this recommended rate.

Insecticide

Malathion is the only insecticide registered for application to grain by farmers in WA. Malathion resistance among insects is widespread throughout the state and you should consider other control methods whenever possible. Malathion will protect your grain for 3-6 months where insects are still susceptible.

Spray the label recommended mix onto the grain at the rate of 1L per tonne as it is augured into the silo. Alternatively, add Malathion dust at 1kg per tonne. Increasing the dose rate will not significantly increase protection. Malathion will break down in 3-6 months under ideal (cool, dry) conditions.

The use of other insecticides for protecting grain is illegal in WA.

GROWING CORN USING PRECISION AGRICULTURE

Scientists in China have been attempting to answer corn producers concerns about precision agriculture, nitrogen fertilizer and their impacts on corn yield, grain quality, resource efficiency, profit and contamination reduction. Precision agriculture is defined as the usage of available technology to develop custom management of soil and crops to fit specific conditions of a small area that is within a larger unit, such as a field.

As of late, many uncertainties have been sprouting up in corn production. Researchers and producers have been wondering if precision agricultural technologies can improve crop yield and quality or reduce their variability. Farmers have been asking a number of questions from, which hybrid should I plant for best yield and quality, to does applying nitrogen fertilizer at a uniform rate produce a better crop outcome, and if not, what nitrogen fertilization strategy does produce a better crop in yield and quality?

Scientists at China Agricultural University, the Precision Agriculture Center of University of Minnesota and Mosaic Crop Nutrition have been attempting to answer those questions by investigating the potential impact of precision nitrogen management on corn yield, protein content and test weight in a new study.

Precision agriculture is defined as the usage of available technology to develop custom management of soil and crops to fit specific conditions of a small area that is within a larger unit, such as a field. This practice has revolutionized modern farming by allowing farmers to choose the best management strategy at a specific time and place in their fields. It has the potential to increase agricultural resource use efficiency, reduce environmental contamination, and maintain or increase crop yield. Corn farmers use this application by varying the rate of fertilizer depending on differences in potential crop yield, soil type and landscape features across the field. As grain markets shift to a greater emphasis on ethanol, more attention is being directed to optimizing grain quality, where traditionally the emphasis was on quantity.

The significant variability of abundance in a given area and abundance over a period of time in crop yield and grain quality has not influenced use efficiency or profit of products made from the crops, but made it difficult for farmers to get premium prices for their products.

The study was conducted on two commercial corn fields in eastern Illinois in 2001 and 2003 involving two corn hybrids and five different N fertilizer application rates across the landscape. Nitrogen response of corn yield and quality were fitted at different within-field locations, and the potential impacts of different N management strategies were evaluated against a uniform rate of N application that is a common farmer's practice in the region.

The results indicated that one hybrid was found to have higher yield, quality and distribution to suppliers than the other hybrid under either a uniform or varied nitrogen application. Results also showed that varying nitrogen applied to localized within-field conditions and hybrid differences could either increase corn yield with similar or higher nitrogen rates or maintain yield with less nitrogen application, without any significant improvement of grain quality.

In many developing countries, there is little to no use of western precision agriculture technology. This is due to smaller field sizes, limited access to technology, financial capital and training. Nevertheless, farmers explore the means and resources available to them in order to increase agricultural productivity and production. Through the prudent and targeted application of inputs, precision agriculture contributes to Sustainable Intensification by enabling farmers to increase their yields with fewer inputs than other application methods such as broadcasting (scattering over a large area) fertilisers or seed, for example. This can also improve soil quality and moisture whilst minimising the environmental impact that excessive input use can cause. Further, the targeted application of inputs can help farmers to be more competitive by lowering production costs. Precision agriculture involves measures such the targeted use of inputs through microdosing, soil testing, and proper seed spacing as opposed to broadcasting.

Yield-mapping technologies

The primary technological advance that made precision agriculture feasible is the yield map, which enables the farmer to estimate crop yields for sections as small as a few square yards and to display the collection of these estimates in color-coded maps. Any area can be mapped. Typical fields are about 150 acres. Growers can identify high- and low-yielding regions of the field and precisely quantify the differences between them. Yield mapping is based on three basic technologies: yield monitors, GPS and GIS.

YIELD MONITOR

A yield monitor is a device that periodically (generally about once per second) measures the mass or flow rate of harvested material, and based on this measurement computes an estimate of crop yield. In a combine harvester, the estimate is obtained by measuring the force of the grain against a plate.

GPS.

The global positioning system, or GPS, uses triangulation of signals from a constellation of satellites to identify the location of the GPS on the Earth's surface, generally within about 1 yard. A fully functional yield-monitoring system includes a GPS that tags each yield estimate with the current location in the field so the data can be matched with the location. These data are stored in a file that can be downloaded after harvest.

GIS.

A geographic information system, or GIS, is a computer program that combines database-management systems with graphics. It can accept data from an assigned location and generate a thematic map showing the spatial distribution of the data. Data from a yield monitor is downloaded into a GIS and converted into a colorcoded yield map that displays yield levels throughout the field. A typical yield monitor includes a data card to transfer files from a personal computer. The typical yield map shows high-yield areas in green and low-yield areas in red.

We demonstrated state-of-the-art precision agriculture technology in a 70-acre wheat field located in Winters, near Davis. The result was a series of interpretive maps that the grower then used to adjust fertilizer and crop distribution during the 1997 growing season. The research was a joint project of UC Davis scientists and Precision Farming Enterprises, a Davis-based consulting firm.

Additional information sources

Soil sampling.

If the yield map reveals substantial variability at different places in the field, as is the case in figure 1, additional information is needed to determine why. Intensive soil samples are a valuable, albeit expensive, source of information. In this procedure, soil samples are collected at a large number of points across the field (typically an average of one per acre or two), analyzed and statistically interpolated to provide an estimate of the soil's physical and chemical properties at every point in the field.

HARNESSING THE MARKET FOR LASTING CHANGE

A cornerstone of this new strategy is equipping farmers with the tools that work. Farmers face mounting regulatory and market pressure to make their operations more sustainable, and technology is creating new economic and environmental opportunities for agriculture.

EDF research has shown that precision nutrient management – paired with cover crops, natural buffers and wetlands – is key to reducing the environmental impact of fertilizer. A new wave of data analytic tools and internet-of-things devices make it easier to achieve that precision.

Because of the great expense associated with this process, other sources of information such as remote-sensing data and soil electrical-conductivity measurements are often used to stratify the field for purposes of improving the statistical efficiency of soil sampling, as well as to supplement or replace the soil samples themselves. In the Winters field, we took a soil sample per acre. Therefore, the low yield in the field's northern portion can be associated with high moisture or poor drainage.

Remote sensing.

Remote sensing is a less expensive source of information than soil monitoring, which can also estimate crop vegetation status. At present, the most common form of remote sensing is via electromagnetic radiation, which is detected by a sensor mounted in an airplane or satellite (also called passive sensing systems). Electromagnetic radiation is classified by wavelength. The wavelengths currently most widely used for agricultural applications are visible light, near infrared and thermal infrared. Plants and soil reflect visible light and near-infrared radiation in relative proportions that provide information about the extent and health of vegetation. Thermal-infrared radiation is emitted from plants at wavelengths dependent on their temperature. Because plants suffering from water or other stress tend to have higher temperatures, the wavelength of thermal-infrared radiation indicates the plant's stress level.

False color infrared shows infrared radiation as red, red light as green and green light as blue. The first was taken in December 1995, shortly after planting, and shows the bare soil following a heavy rainfall. The darker areas are wetter and correspond to the heavy clay with lower infiltration rate. The second was taken in March 1996 at the soft-dough stage, when the wheat grains were immature. The brighter red areas in the southern end of the field indicate healthy vegetation, while the grayer areas in the northern portion indicate less vigorous vegetation and a less dense canopy, presumably caused by aeration stress. The third image was taken in May 1996, just prior to harvest. Some of the brighter red areas in this figure correspond to weedy patches.

Nutrient Star shows what works, where

So in 2016, EDF launched NutrientStar, an independent, science-based program that rigorously tests and reviews the performance of commercially available nutrient management tools. It's like a 'Consumer Reports for agriculture,' with results that are freely available for anyone to use.

NutrientStar gives farmers an alternative to relying on marketing materials alone. It lets them know which tools have been tested on which crops, where each tool was tested and what the impact was on yield, fertilizer efficiency and profitability. An independent review panel of respected experts keeps the results objective.

Advances in software and data analytics now allow farmers to calculate costs on every square foot of their farms with unprecedented accuracy, unlocking potential solutions for pest issues, soil health issues and more.

With more data sources coming online, NutrientStar promises to become even more powerful as a go-to resource for crop advisors in their work with farmers.

Food suppliers take solutions to scale

To make precision agriculture techniques standard practice, EDF has been building relationships with food suppliers, providing expertise to help them source sustainably produced agricultural products.

Electrical-conductivity measurement.

A second relatively inexpensive source of field information is electrical-conductivity measurement. Soil electrical conductivity can be measured either directly or through measurement of electrical inductance. In either case, the machine measures the strength of electric currents passed through the soil. Measurements are generally correlated with soil salinity, water content, soil density, topsoil depth and clay content. Figure 4 uses colors to show soil electrical conductivity in the Winters wheat field. In this low-salinity soil, electrical conductivity is related to clay content.

Additional data can be obtained from scouting of insect, weed and pathogen infestations. In field scouting, a trained person walks through the field to evaluate weed levels. This map was developed based on a combination of scouting on a one per acre grid and interpretation of the May 1996 aerial photo.

Variable-rate technology.

Maps of data from soil analyses, remote sensing and direct measurements of electrical conductivity and pest and pathogen levels can be combined with yield-map data in the GIS to determine the primary factors underlying yield variability. If these factors can be determined and the decision is made to vary the application rates of pesticides and fertilizers across the field, variable-rate technology (VRT) application equipment can be utilized.

This equipment typically contains an on-board GPS to determine precise locations in the field, as well as adjustable-rate application devices. These may be either manually adjusted by the operator or equipped with an on-board computer that adjusts the rate automatically. VRT closes the loop between acquiring the data, converting the data into information, and acting on the information to make maximum use of underexploited areas of the field and reduce wasted inputs in low-potential areas.

Based on the information provided in this figure, the cooperating grower carried out two modifications to his management practices; he changed his fertilization program to increase available phosphorous and broke off the southern portion of the field to grow a higher-value vegetable crop. The grower is also exploring the economics of installing drain tiles in the field's northern section to improve drainage.

ENVIRONMENTAL, REGULATORY ISSUES

In California, agriculture is regulated by several state and federal agencies. In 1990, California was the first state in the nation to require 100% reporting of all pesticides applied and to compile and publish an annual database of the amounts of all chemicals used in agriculture. New agricultural chemicals must be registered with the U.S. Environmental Protection Agency as well as the California Department of Pesticide Regulation. The intent of California's strict compliance program for pesticides is to reduce non-point-source water pollution related to agriculture.

U.S. EPA is conducting studies in several watersheds to determine mechanisms to reduce non-point-source pollution. One possible solution would be to regulate fertilizer inputs so that the agricultural producer is allowed a certain number of fertilizer units per farm, based on soil type, crops and location. To maximize farm productivity, the inputs could be variably spread across each field depending on soil type and future crops.

Current regulations do not mandate a site-specific fertilization plan, but such plans may be specified in the future. If such plans are mandated, growers will need to more closely match input application rates with crop demands and reduce unnecessary applications to areas where no economic benefits are achieved.

In California, water is a critical resource in the arid valleys. Over many decades, state and federal agencies have created an enormous reservoir system to satisfy California's agriculture industry. Agriculture's allocation of approximately 75% to 80% of California's stored water must compete with the state's growing urban population and a renewed interest in the environment. Agriculture has taken significant strides to understand each crop's water needs and, with the adoption of drip irrigation, to alter water usage in most permanent tree and vine crops.

Adoption of precision agriculture

Precision agriculture is in the early adoption stage; the technology is commercially available but not yet widely used. The technology is currently most advanced in combine-harvested grain crops. In 1998, an estimated 17,000 combines in the United States were equipped with yield monitors (Anonymous 1998). However, this accounted for only 8% of major U.S. crops and 5% of farmers with gross annual incomes above $100,000.

Cochrane (1993) introduced the concept of the "technological treadmill" to characterize the spread of innovations in agriculture. According to this theory, market pressure drives commodity prices toward the minimum average cost of production. Early adopters of a technology that increases profitability will benefit during the period when prices still reflect the production costs of those who have not adopted it. As more farmers either adopt the technology or are driven out of business, the average cost is driven down to that of users of the new technology. Early adopters may be motivated by reasons other than economics (for example, curiosity or willingness to speculate on innovation), but the ultimate widespread adoption of a technology depends on its ability to increase profitability.

At the national level, studies on the profitability of precision agriculture have been inconclusive (Lowenberg-DeBoer and Swinton 1997). In part this is because its benefits are often hard to isolate and quantify. SSM has both tangible and intangible associated costs. Tangible costs include equipment, such as the yield monitor, GPS and GIS, and services, such as remote sensing, soil profiling and data management. Intangible costs, such as the time and effort required to learn a new concept like SSM and make it a seamless part of the overall farming operation, are more difficult to quantify.

Because precision agriculture involves the use of information, its profitability can be increased by making the information cheaper and easier to acquire and by increasing the effectiveness of the information once it is acquired. A well-established characteristic of information technology is that costs are reduced dramatically as the technology matures. In the long run, the primary restriction on adoption of precision agriculture technology is not likely to be cost but rather the ability of the farmer or consultant to make effective use of it to increase profits. Experience with other information technologies indicates that some time is required for users to learn how to employ these technologies effectively.

Eventually, it is likely that, as they have with other new technologies, growers will adapt the information technologies to their production systems.

By the mid-21st century something akin to the scenario is likely to be in place on farms throughout California. Planting, fertilization, pest treatment and harvesting will all be connected to centralized data-recording and control systems. Because one of the key technological bottlenecks in precision agriculture at the moment is the cost of crop and soil information, sensor technology must be dramatically improved. We can expect to see implement-mounted sensors that use spectrography to detect soil properties such as mineral nutrient concentration, salinity and sodicity. Similarly, detailed analysis of the electromagnetic spectrum reflected and emitted by plants will be used to interpret water status, pest pressure and nutrient level. Preliminary research is currently under way, and will certainly have reached fruition by midcentury, to genetically engineer plants that provide a direct indication of their status through changes in their reflectance or emittance properties.

Future in focus: Site-specific management

We cannot say with any certainty that the scenario just described will ever come to pass, or that any aspect of precision agriculture will survive the early adoption process and achieve widespread use. It is likely, however, that some aspects will indeed be adopted. The cost of newly introduced information technologies has historically declined dramatically as these technologies matured. This is likely to happen with the technologies associated with precision agriculture, particularly yield monitors, sensors and GPS.

Likewise, the value of the information is likely to increase as farmers become more familiar with it. Nowak (1997) distinguishes between "soft" and "hard" precision agriculture technologies. "Soft" technologies rely more on traditional means of acquiring site-specific information (for example, visual observation of crop and soil differences) and on farmer intuition rather than statistical or scientific analysis of data. Because the cost of the information is likely to decline, farmers are likely to increasingly rely on "hard" technology to acquire information. On the other hand, if history is any guide, farmers are more likely to rely on "soft" technology to interpret map data. However, "hard" devices such as on-the-go weed-control systems based on the automated detection of weeds through video imaging may also come into widespread use.

Several critical issues will come into play with the adoption of precision agriculture: % Size-neutrality.

Many of the technological advancements associated with agricultural mechanization have favored large farms, resulting in a trend toward farm consolidation. Because SSM may increase the efficiency of large-scale crop production, it is reasonable to suppose that it may have the same effect.

Farm jobs.

In general, the incorporation of information technology in industry has tended to expand the market for skilled employees and reduce the number of semiskilled and unskilled jobs. In agriculture, one may expect SSM to increase the demand for skilled individuals who can manage and interpret the information. This increase in skilled positions may be associated with a corresponding decrease in the number of available unskilled positions.

FROM LOCAL DATA TO GLOBAL IMPACT

EDF's approach helps craft markets in which everyone has an economic incentive to protect the environment, and this approach is reaping rewards.

We're already halfway to our 2022 goal of securing commitments to improve fertilizer and soil management practices on 45 million acres, half of U.S. corn acreage – a tipping point that should accelerate further gains. And we're working with our partners to track progress toward implementing those commitments.

With 40% of the world's land used for food production, wider adoption of these practices could improve water quality and reduce net carbon emissions on a worldwide scale, a big win for everyone.

Unintended side effects.

Regulatory agencies may require site-specific management plans to reduce the environmental impacts of agricultural inputs. This information may influence the price of land in the way that laser leveling of fields does now. It is possible that farmers' ability to more precisely measure variability and discern cause and effect will encourage on-farm experimentation, changing the relationship of farmer and university researcher by making the farmer a part of the entire research program rather than a passive recipient of information.

Finally, precision agriculture may have other uses that are totally unanticipated by those conducting research on the technology today. A fundamental role of UC researchers will be to adapt novel information technologies to improve productivity and reduce environmental effects of crop production.

The current state and the future goals

Although precision agriculture technologies and techniques are extremely positive and potentially revolutionary, the innovations and inventions of precision agriculture are widely spread yet as they should be, largely attributed to the high costs involved.

Moreover, there is also a lack of the necessary skills and experience required to utilize best the variety of tools and a huge amount of data that is collected by the same technologies. The end game for precision agriculture is to develop plans that are very flexible and adaptable to changes that are based on the collected information so as to improve farm productivity. However, for precision farming to be established properly and change or transform the current state of traditional agricultural practices, it is important that focus on figureheads of the industry today, that is, the farmers, must be given the required support and muscles to adopt and implement this new way of farming.

On the other hand, there will be need to rid out the old stereotypes, by being farmers up-to-date and training them as a preparation for adopting new farming techniques. Thus the new and older generation of farmers should be encouraged – especially the younger generation.

Technologies such combination of machine learning and image recognition in applying the exact amounts of fertilizer, seeds or even during irrigation would result in reduced need for adjuvants. In specialty plants, precision farming may include field Internet-of-Things and highly defined camera connectivity being used to showcase the return on investment of biological products that include mating disruptors and pheromones. The end result of this would be the adoption of biological products and forego the traditional chemicals.

With the years of focus on alternative ways of gathering field data like satellites, drones, field sensors, robots and so forth, which are integrated with analyzing and synthesizing tools and software, considerable attention has now shifted to variable rate application (VRA) of crop inputs. A study that was conducted by Purdue University regarding precision services offered by precision farming retailers indicated that variable rate application of fertilizer, lime, and seed has boomed over the last several years. At the same time, variable rate application of farm chemical use has reduced.

The management of agricultural fields at spatial scales that are smaller than an individual field is the future of sufficient production. The increased population and squeezed natural resources would require better farming techniques. At its most basic level, precision farming should be based on information management, and possibly made by a range of new technological developments. This would provide increased profitability opportunities and reduced environmental damage caused by farming activities. Hopefully, government, farmers and other relevant stakeholders would come to continue to appreciate the economic and environmental benefits that come along with precision agriculture.

PRECISION FARMING COULD HELP REDUCE CLIMATE GAS EMISSIONS

Improved fertiliser, soil and water management can significantly reduce climate gas emissions, while maintaining similar yields and reducing production costs: a typical win-win. However, reports show low levels of adoption. What can be done to mainstream precision farming?

Agriculture is very much affected by the effects of climate change: more extreme and changing patterns in temperatures and rainfall require agricultural practices to adapt.

Precision farming reduces greenhouse gas emission

In a recent study, all available literature on precision agriculture was reviewed regarding the impact of precision agricultural measures on:

- climate
- environment
- farm economy

The review revealed that most scientists were focussing on productivity and the economic benefit of precision agriculture, the environmental and climate mitigation effects were often only mentioned through reduction of inputs achieved.

EFFECT OF PRECISION FARMING TECHNOLOGIES ON CLIMATE

Recently, more scientific attention has been given specifically to the climate effect of precision agriculture technologies. The study identified that precision agriculture techniques have a positive contribution to greenhouse gas reduction.

Nutrient and water management have a direct contribution, in particular the so-called Variable Rate Technology for nutrients application because of the lowered N2O emissions. Although results vary between soils and weather conditions during application, variable rate application can reduce fertiliser use by 10-30%, while maintaining the same productivity. Besides the reduction in costs, there is a very important contribution to reducing greenhouse gas emissions.

All farmers should adopt precision agriculture technology

Agriculture is accountable for 10-15% of all human induced climate gas emissions worldwide. Arable farming emits CO2 and N2O. N2O is a 300 times stronger climate gas then CO2. In the Paris Agreements, many governments made commitments to significantly reduce emissions, to slow down the global warming process. Meeting the climate goals of the Paris Agreements requires many, preferably all, farmers to work with precision agriculture technology.

One limitation of the papers reviewed is that only a few actually measured directly environmental indices, such as leaching with the use of soil sensors. Most of them estimated indirectly the environmental benefits by measuring the reduced chemical loading. Results from an on-farm trial in Argentina provide an example of how site-specific information and variable rate application could be used in maintaining profitability while reducing N applications. Results of the sensitivity analysis show that PA is a modestly more profitable alternative than whole field management, for a wide range of restrictions on N application levels. These restrictions might be government regulations or the landowner's understanding of environmental stewardship. In the example, variable rate of N maintains farm profitability even when nitrogen is restricted to less than half of the recommended uniform rate.

EU data on why smart farming is not adopted is scarce

The European Commission could think of stimulating the uptake for instance by rewarding smart farmers through the CAP. This could reach all farmers in the EU and thus provide a huge contribution to a reduction of greenhouse gas emissions as well as preserving soil organic carbon. Unfortunately, neither adoption levels nor the motives and barriers of farmers to adopt are known. Structural surveys measuring the adoption of precision agriculture are scarce. In the United States, the Purdue University makes an annual assessment of sales and usage of precision agriculture. This US study also collects opinions of farmers, dealers and contractors. But similar structural adoption surveys are missed in Europe.

CAPIGI PAYS SPECIAL ATTENTION TO CLIMATE SMART FARMING

To conclude, an increase in farms using precision agriculture techniques will have a huge impact on the reduction of greenhouse gas emissions and hence on climate change. Scientists should focus more on the climate mitigation potential. The machine manufacturers, a main supplier of precision agriculture technologies, can promote this by reducing the complexity of precision agriculture for instance by better implementing sector wide standards, harmonising their marketing messages and focus on ease-of-use.

PRECISION AGRICULTURE: BASICS OF GPS

The development and implementation of precision agriculture or site-specific farming has been made possible by combining the Global Positioning System (GPS) and geographic information systems (GIS). These technologies enable the coupling of real-time data collection with accurate position information, leading to the efficient manipulation and analysis of large amounts of geospatial data. GPS-based applications in precision farming are being used for farm planning, field mapping, soil sampling, tractor guidance, crop scouting, variable rate applications, and yield mapping. GPS allows farmers to work during low visibility field conditions such as rain, dust, fog, and darkness.

In the past, it was difficult for farmers to correlate production techniques and crop yields with land variability. This limited their ability to develop the most effective soil/plant treatment strategies that could have enhanced their production. Today, more precise application of pesticides, herbicides, and fertilizers, and better control of the dispersion of those chemicals are possible through precision agriculture, thus reducing expenses, producing a higher yield, and creating a more environmentally friendly farm.

Precision agriculture is now changing the way farmers and agribusinesses view the land from which they reap their profits. Precision agriculture is about collecting timely geospatial information on soil-plant-animal requirements and prescribing and applying site-specific treatments to increase agricultural production and protect the environment. Where farmers may have once treated their fields uniformly, they are now seeing benefits from micromanaging their fields. Precision agriculture is gaining in popularity largely due to the introduction of high technology tools into the agricultural community that are more accurate, cost effective, and user friendly. Many of the new innovations rely on the integration of on-board computers, data collection sensors, and GPS time and position reference systems.

Many believe that the benefits of precision agriculture can only be realized on large farms with huge capital investments and experience with information technologies. Such is not the case. There are inexpensive and easy-to-use methods and techniques that can be developed for use by all farmers. Through the use of GPS, GIS, and remote sensing, information needed for improving land and water use can be collected. Farmers can achieve additional benefits by combining better utilization of fertilizers and other soil amendments, determining the economic threshold for treating pest and weed infestations, and protecting the natural resources for future use.

GPS equipment manufacturers have developed several tools to help farmers and agribusinesses become more productive and efficient in their precision farming activities. Today, many farmers use GPS-derived products to enhance operations in their farming businesses. Location information is collected by GPS receivers for mapping field boundaries, roads, irrigation systems, and problem areas in crops such as weeds or disease. The accuracy of GPS allows farmers to create farm maps with precise acreage for field areas, road locations and distances between points of interest. GPS allows farmers to accurately navigate to specific locations in the field, year after year, to collect soil samples or monitor crop conditions.

Crop advisors use rugged data collection devices with GPS for accurate positioning to map pest, insect, and weed infestations in the field. Pest problem areas in crops can be pinpointed and mapped for future management decisions and input recommendations. The same field data can also be used by aircraft sprayers, enabling accurate swathing of fields without use of human "flaggers" to guide them. Crop dusters equipped with GPS are able to fly accurate swaths over the field, applying chemicals only where needed, minimizing chemical drift, reducing the amount of chemicals needed, thereby benefiting the environment. GPS also allows pilots to provide farmers with accurate maps.

SPATIAL LOCATION

The origin of Precision Agriculture begins with researchers collecting soil samples, and using spatial statistics methods to determine the different soil types in a field. From this analysis, the researchers developed soil maps. Farms were early adopters of both GPS and Geographic Information Systems (GIS) technologies. As civilian GPS became more accurate, farms started to utilize this technology to increase the accuracy of operational spatial data. Collecting spatial data from equipment and sensors that allowed farms to pinpoint the high yield areas. Also, using GPS data to determine where to increase or decrease pesticides, fertilizers use and irrigation.

SOFTWARE AS A FARM INPUT

The business case for Precision Agriculture methods is the reduction of farm inputs. Farm inputs range from equipment, seeds, pesticides to energy. In turn a reduction in farm inputs reduces the overall cost to produce a harvest. Modern farm inputs include an Internet of Things (IoT) ecosystem - networks of advanced sensors, auto guided driving tractors (harvesting both crops and data), servers and drones. Furthermore, data purchased is also a farm input. The data purchased by the farm assist in the planning and execution of the harvest, such as weather feeds, and satellite imagery.

Harvesting, Both Crops and Data

During planting and harvest, these farm inputs yield more data for the farm to analyze. Location of tractors and their status, drone flights to find areas of crops under duress and the yields of individual fields. Smaller farms may only generate a few hundred rows of data a day, while a large farm could generate billions.

According to Business Insider, the amount of data generated by the farm will dramatically increase over the next 20 years.

OPEN SOURCE SOFTWARE, AND FARMING

Free and Open Source Software for Geospatial (FOSS4G) technology can play a role in reducing the cost of software inputs, and allow farms to analyze and visualize their data.

Two LocationTech software projects, hosted at the Eclipse Foundation, can assist farms with the analysis of large amounts of spatial data: LocationTech GeoMesa and LocationTech GeoTrellis.

GeoMesa

GeoMesa brings spatiotemporal analytics and visualizations to Accumulo, a scalable data storage and retrieval system. Farms that leverage GeoMesa will be able to perform large scale analysis on their vector data. Vector data is normally collected by sensors and tractors - general operational data. Vector data can also include weather reports, and any data that has been converted from raster to vector.

By giving the farm the ability to process these massive operational data sets, insights into the impact of the day to day operations on the harvest become clearer. These analyses will give the farm a better understanding of how to manage operations and apply these lessons learned to the next harvest season. To learn more about GeoMesa, read this article.

GeoTrellis

GeoTrellis can batch process massive amounts of raster data. With daily satellite updates, and on the fly drone collection, the amount of raster data collected by farms has never been greater.

By using GeoTrellis a farm can analyze vast amounts of raster data at a low cost. This will help the farm identify where plants are under duress or which areas are being over or under irrigated. Through the processing of this raster data, the farm will better understand trouble spots, and can investigate those targeted areas. Smaller farms can also benefit by using FOSS4G technology. Farms with a web based GIS can use Geoserver to generate and maintain web maps and applications. Postgres/PostGIS can provide spatial data management and geoanalytics of farm data.

For day to day desktop spatial analysis, a farm could use QGIS or gvSIG. Both of these desktop applications can perform robust geospatial analytics and generate maps. By reducing the costs of their software inputs to perform both large and small scale geoanalytics and visualizations, the farm will be able to increase the performance of their operations and increase efficiencies.

TECHNOLOGY TRANSITION

Precision agriculture is dependent on the existence of variability in either or both product quantity and quality. If this variability does not exist then a uniform management system is both the cheapest and most effective management strategy and precision farming is redundant. Thus, in precision farming, "Variability of production and quality equals opportunity".

Having said this, the nature of the variation is also important in determining the potential for PA in a system. For example the magnitude of the variability may be too small to be economically feasible to manage. Alternatively the variability may be highly randomized across the production system making it impossible to manage with current technology. Finally the variability may due to a constraint that is not manageable. Thus the implementation of precision farming is limited by the ability of current variable rate technology (VRTmachinery/technology that allows for differential management of a production system) to cope with the highly variable sites and the economic inability to produce returns from sites with low variability using precision farming (VRT). Due to these constraints PA is at present operating on a zonal rather than a completely site-specific basis. As VRT improves and the capital cost of entering PA decreases, the minimum size of management zone needed to effectively implement PA will decrease till eventually a truly site-specific management regime is possible.

Until this occurs there is need to be able to quantify both the variability of a production system and the size of the minimum manageable zone (MMZ). If the variability in the production system dictates management zones smaller than the MMZ than PA is not relevant to the system. At the present time (but may be in future). It will be interesting to see how the concept of the management zone develops and to see how it compares with the concept of terroir.

BIBLIOGRAPHY

- Agricultural engineering Britannica Online. Retrieved 2012-12-25.
- Tractor (vehicle) Britannica Online. Retrieved 2012-12-25.
- "How to Manage Prevent Plant Weeds". Successful Farming. 2019-12-06. Retrieved 2020-08-13.
- "New High-Tech Farm Equipment Is a Nightmare for Farmers". Wired. 2015-02-05.
- "We Can't Let John Deere Destroy the Very Idea of Ownership". Wired. 2015-04-21.
- Exemption to Prohibition on Circumvention of Copyright Protection Systems for Access Control Technologies http://copyright.gov/1201/2015/fedreg-publicinspectionFR.pdf
- Farmlabs.org. "This is the Beginning of a network of open laboratories for agricultural research and experimentation". farmlabs.org. Retrieved 2019-07-20.
- Gaillard, Chris. "L'Atelier Paysan". L'Atelier Paysan (in French). Retrieved 2019-07-20.
- Chance, Quentin; Meyer, Morgan (2017-06-06). "L'agriculture libre. Les outils agricoles à l'épreuve de l'open source". Techniques & Culture. Revue semestrielle d'anthropologie des techniques (in French) (67): 236–239. doi:10.4000/tc.8534. ISSN 0248-6016.
- "Ekylibre". ekylibre.com. Retrieved 2019-07-20.
- "Group Overview ‹ Open Agriculture (OpenAg)". MIT Media Lab. Retrieved 2019-07-20.
- "Project Overview ‹ Open Phenome Project". MIT Media Lab. Retrieved 2019-07-20.
- "recipe:start [OpenAg]". wiki.openag.media.mit.edu. Retrieved 2019-07-20.

- Jan H. Maas (1994) De Nederlandse agrosector: geografie en dynamiek p.177
- Gifford (1992, p. 9) even stated that "nearly all farm machinery manufacturers today can be trace their beginnings to a blacksmith."
- Ganzel, Bill (2003). "Tractors". Farming in the 1930s. Wessels Living History Farm.
- Collins, Edward JT. "The rationality of'surplus' agricultural labour: mechanization in English agriculture in the nineteenth century." The Agricultural History Review (1987): 36-46.
- Dewey, Peter. "The British Agricultural Machinery Industry, 1914-1939: Boom, Crisis, and Response." Agricultural History (1995): 298-313.
- McNeil, Ian (1990). An Encyclopedia of the History of Technology. London: Routledge. ISBN 0-415-14792-1.
- Hounshell, David A. (1984), From the American System to Mass Production, 1800–1932: The Development of Manufacturing Technology in the United States, Baltimore, Maryland: Johns Hopkins University Press, ISBN 978-0-8018-2975-8, LCCN 83016269, OCLC 1104810110
- Wells, David A. (1891). Recent Economic Changes and Their Effect on Production and Distribution of Wealth and Well-Being of Society. New York: D. Appleton and Co. ISBN 0-543-72474-3. RECENT ECONOMIC CHANGES AND THEIR EFFECT ON DISTRIBUTION OF WEALTH AND WELL BEING OF SOCIETY WELLS.
- Ford, Henry; Samuel (1922). My Life and Work: An autobiography of Henry Ford.
- Constable, George; Somerville, Bob (2003). A Century of Innovation: Twenty Engineering Achievements That Transformed Our Lives, Chapter 7, Agricultural Mechanization. Washington, DC: Joseph Henry Press. ISBN 0-309-08908-5. [page needed]
- White, William J. "Economic History of Tractors in the United States". Archived from the original on 2013-10-24.
- Ayres, R. U.; Ayres, L. W.; Warr, B. (2002). "Exergy, Power and Work in the U. S. Economy 1900-1998, Insead's Center For the Management of Environmental Resources, 2002/52/EPS/CMER" (PDF). Archived from the original (PDF) on 2017-08-13. Retrieved 2010-10-06.
- Moore, Stephen; Simon, Julian (Dec 15, 1999). "The Greatest Century That Ever Was: 25 Miraculous Trends of the last 100 Years" (PDF). The Cato Institute: Policy Analysis, No. 364. Fig 13.

- Wall Street Journal: "Robots Step Into New Planting, Harvesting Roles - Labor shortage spurs farmers to use robots for handling delicate tasks in the fresh-produce industry" By ILAN BRAT April 23, 2015
- Washington State University Extension - School of Economic Sciences: "Asparagus Production Economics" December 8, 2010
- Vegetable Growers News: "Mechanical Asparagus Harvester Almost a Reality" December 21, 2009
- Fox Business News: "Machinery takes the place of migrants as Maine's blueberry harvest booms" September 06, 2015
- Albuquerque Journal: "Chile harvester gets a field test in NM" By Diana Alba Soular September 22, 2014
- Current Argus: "Editorial: Trial could be a boon to chile farmers" Archived 2015-08-11 at the Wayback Machine July 31, 2015
- KJZZ: "New Mexico Researchers To Test Mechanized Green Chile Harvesting" by Carrie Jung July 29, 2015
- Las Cruces Sun: "Experts: Machines could reverse declining New Mexico green chile acreage" By Diana Alba Soular Archived 2015-08-10 at the Wayback Machine July 25, 2015
- University of California Davis Migration Files: "The Status of Labor-saving Mechanization in Fruits and Vegetables" By Wallace E. Huffman May 25, 2010
- Farm 2 ranch magazine: "Labor concerns lead peach growers to look to machines" August 18, 2016
- U. S. Produce Industry and Labor: Facing the Future in a Global Industry By Linda Calvin retrieved September 28, 2013
- Fresno Bee: "New raisin grape holds promise for central San Joaquin Valley growers" By Robert Rodriguez September 19, 2015
- Japan Times: "Latest robot can pick strawberry fields forever" September 26, 2013
- "No Hands Touch the Land: Automating California Farms" (PDF). California Agrarian Action Project: 20–28. July 1977. Retrieved 2015-04-25.
- Keith Cowling, David Metcalf, A. J. Rayner. (1973) Resource Structure of Agriculture: An Economic Analysis, p. 127-8
- World bank (1954) Western Europe - Tractor and agricultural machinery industry.
- VDMA. Agricultural Machinery November 2013 : Market Perspectives 2014[permanent dead link], 2013-14. p. 5
- Ade Asefeso (2012). CEO Guide to Doing Business in Australia, p. 54

- Kenneth James, Narongchai Akrasanee. Small and Medium Business Improvement in the ASEAN Region: Production Management. Institute of Southeast Asian Studies, 1 dec. 1988
- The leading machinery manufacturers worldwide in 2012, based on revenue. Accessed 25.11.2014.
- Number of employees of John Deere (Deere & Company) from 2002 to 2013. Accessed 25.11.2014.
- Revenue of tractor & agricultural machinery manufacturing (NAICS 33311) in the United States from 2009 to 2014 (in billion U.S. dollars). Accessed 25.11.2014.
- Yuan'en, Gao (2007). "Agricultural Machinery Industry in China" (PDF). Chinese Association of Manufacturers of Agricultural Machinery.
- Global and Chinese Agricultural Machinery Industry Report, 2013-2014, December 2013. Summary (online). Accessed 25.11.2014.
- Singh, Gyanendra (2006). "Agricultural Machinery Industry in India (Manufacturing, marketing and mechanization promotion)" (PDF). Status of Farm Mechanization in India. Archived from the original (PDF) on March 18, 2013.
- Singh, Sukhpal (April 2009). "Agricultural Machinery Industry in India: A Study of Growth, Market Structure, and Business Strategies" (PDF). Indian Institute of Management Ahmedabad. Archived from the original (PDF) on 2015-09-24.
- "Agricultural Machinery Industry in Turkey". blog.tcp.gov.tr. Turkish Ministry of Economy. 1 June 2011. Archived from the original on December 5, 2014.

INDEX

A

Absorption 15

Adopters 271

Adoption 212, 231, 244-245, 254, 271-273, 276

Aeration 264-265

Aerial 233

Agronomics 230-231, 244-245

Agronomy 241

Air-cooled 79

Atmosphere 78

B

Bacteria 262

Bamboo 186

Barriers 21

Biodiesel 23

Biofuels 12, 23

Biogas 12

Biomass 4, 22-23

Biopower 23

Bulldozers 192, 196

Bullocks 2, 158, 169

Business 220, 249, 278, 283-284

C

Cameras 257

Campbell 248

Canals 12

Capacitor 117

Carbon 88, 90, 167, 184, 198

Carburetor 75

Cellular 249

Channels 159

Chemicals 41, 202, 223, 270, 277

Chemistry 37

Chisel 142, 144

Circuits 120

Cleaners 213

Climate 6, 29, 275-276

Clutch 94, 108, 111-114

Combustion 3, 27, 35, 55-57, 60, 67, 77, 91, 190, 199, 205

Components 20, 60, 78, 93, 128

Conversion 40, 43

Coolant 79

Copper 117

Cotton 44, 49, 176

Craftsmen 212

Crankshaft 59, 63-65, 67-68, 71, 90, 93

Cultivator 50, 153-156

Cylinders 58, 60, 78

D

Demand 11-12, 33, 36, 209

Diesel 3, 10, 15, 40, 57-58, 75-76, 111, 201, 211, 213

Digital 241

Domestic 197

Drones 227, 230, 238, 257
Dynamics 219
Dynamite 39

E

Economy 45
Emissions 275-276
Energy 1, 3-7, 10-15, 17, 22-25, 38-40, 42, 55, 64, 75, 115, 200, 210-211
Explosion 67

F

Family 220, 222
Farmers 10, 17, 22-23, 29, 33, 175, 198, 200, 207-209, 214-217, 227, 231, 233-238, 240-241, 245-246, 251-254, 257, 260-261, 266-269, 271-272, 274-277, 283
Fertiliser 275
Fertility 243
Fertilize 51
Fertilizer 51, 163, 166-167, 170, 228, 231-232, 235-236, 239, 244, 246, 250-251, 254, 258-259, 266, 268, 271, 274
Flexible 150, 224
Flywheel 63-64, 67
Fossil 6
Framework 178, 218
Fruits 13-14, 16, 185, 258
Fumigation 263, 265
Furrows 125, 137, 162, 166, 170

G

Gasoline 189, 199
Gearboxes 197
Generation 274
Geothermal 23
Graders 213
Grains 261
Gravity 191
Greenhouse 24, 275-276
Growers 35, 245, 248, 258, 283
Growth 9, 40, 208, 211-212, 250, 256

H

Harrows 50, 146, 148-150, 207
Harvest 13, 35-36, 52, 185, 221-222, 224-225, 240, 250, 278-279
Harvesters 34-35, 213
Hazards 198
Herbicides 7, 249
Horsepower 195
Hybrids 224

I

Indicators 233
Infrared 269
Injectors 76
Insects 7, 14, 262-263, 265
Internet 239, 249
Irrigation 8, 11-12, 17-22, 33, 51, 200, 213, 215, 237, 248, 257, 261

L

Landscape 191, 266
Length 97, 99-100, 169, 186
Livestock 257
Loppers 174
Lubricants 87-88

M

Machine 28, 30, 34, 43, 45, 49, 51-53, 103, 108, 115, 119, 162, 170, 177, 184, 190, 196, 203, 239-240, 283
Machinery, 9, 27-28, 30-33, 44, 48-51, 143, 189, 192, 197, 199, 205, 208-209, 212-214, 216-222, 282-284
Magnet 116
Magnetic 117
Models 234

Movement 71, 82, 87

N

Nitrogen 232-233, 250, 266-267

O

Operations 2, 10, 50, 105, 122, 145, 175, 178, 197, 211, 215-217, 221, 223, 279

P

Pathogen 270

Pesticides 7, 11, 49, 206, 246, 270, 278

Petroleum 6

Phosphine 263, 265

Plantation 201

Population 8, 214

Prototype 35

Q

Quality 14, 185, 214-215, 218, 243, 250, 261, 263, 266-267, 279

R

Radiation 269

Radiator 80-85

Reapers 34

Renewable 1, 4, 6, 22

Revolution 27, 34, 38, 44, 207, 231-232, 254-255

Robotics 248

Rodents 14

Rootstock 171

Rotation 117

S

Safety 190, 197-198, 200, 214

Satellites 228, 231, 238, 245

Sensors 227, 233, 239, 247, 249, 253, 263, 272, 278

Simulation 20

Software 29, 231, 241, 245-246, 278

Systems 1, 11, 13, 17-20, 22-24, 29, 41, 46, 228-229, 235, 247-248, 253, 259, 261, 272, 277

T

Tapioca 185

Tension 113

Testing 214

Thermostat 81, 83-84

Threshers 53, 213

Turbines 13, 24, 40

V

Valves 70

Variation 242

Vegetation 269

Viscosity 88

W

Wastage 252

Wavelength 269

Wealth 282

Weather 2, 250

Weeding 10, 175-177

Weevils 262

Wireless 249, 253